12:21:12:11:11:00:00                                    2012

12:21:12:11:11:00:00                    2012

# Dedication

## To ~ The Creator of All That Is...

Thank you Lord God, The Alpha and The Omega, without Whom nothing would ever be possible. Thank you for all of us, our moment of consciousness, and this extraordinary planet and universe in which we live.

I dedicate this book unto You and ask that you Look Upon and protect all of us as we make our ominous passage towards and through our FATED DESTINY. We ask that you keep all of us and this most amazing planet we call home CLOSE to your LOVE and GLORY during this amazing, yet hazardous JOURNEY Home...

AMEN

# Dedication

## To ~ Mr. George Noory

**THANK YOU MR. NOORY**

For Being My LIGHT Amidst the DARKNESS

For Bringing that LIGHT to SO MANY of US on this
Planet

Who – Like Myself– Had No Place to Turn

No Shelter From the Storm.


For Giving Me the Strength to Speak the Truth

For Showing Me that There IS a World out There
Somewhere

That is Open and Hungering to Listen and to Learn
the TRUTH

&

ABOVE ALL

For Giving me the Knowledge That

...Finally...

## I Am NOT ALONE
## There are MANY OTHERS
All of Us "Others"
Who would Never have Known
That We were Not Alone

If Not for the Heart, Soul, Genius, Bravery
and Life-Long Dedication
of ONE MAN
## Mr. George Noory.

12:21:12:11:11:00:00                              2012

**Thank You Mr. Noory**

**For Bringing a Torch of Light**

**Into this Sometimes Dark and Hidden World**!

Mr. George Noory has dedicated his entire life to seeking out and speaking the TRUTH about the world...and the universe...in which we live. He is currently the Number One Talk Show Host on AM Talk Radio with a worldwide audience of millions that reaches into our planet's remotest areas.  George Noory can be heard nightly on: **Coast-to-Coast AM Talk Radio**.

**For information on how to access his show  in your area please visit:**

**Coast-to-CoastAM.com.**

12:21:12:11:11:00:00                                        2012

8

# Dedication
## To My Beauxdreaux

I would like to thank you MY DEAREST, constant
companion ~ my soulmate, my LOVE ~
**My Beauxdreaux**…
for all of your  PATIENCE and LOVE as you sat loyally
by my side during the long, never-ending hours ~
days ~ weeks ~  and months
it took to complete this book.

## To YOU…My Special Reader

Perhaps **YOU**  were **DESTINED** to *"be here now"*
at this amazing, critical time…a time that is
unprecedented throughout the history of mankind
as we know it to be.

12:21:12:11:11:00:00                              2012

# 2012

## Apocalypse Rising

## By: Mayte DeFerrare

ISBN 10:1456350153

ISBN 13: 9781456350154

# L O S T

"MOST BELIEVE THAT WHAT IS DONE..

...IS DONE

YOU CANNOT CHANGE FATE NO MATTER
HOW HARD YOU TRY.

AND

THOSE THAT CHALLENGE WHAT HAS BEEN
DESTINED WILL ALWAYS BE LEFT WITH
DISAPPOINTMENT.

FOR FATE HAS A WAY OF CHARTING

IT'S OWN COURSE...

....BUT....

BEFORE ONE SURRENDERS to the HANDS of
DESTINY

ONE MIGHT CONSIDER

THE POWER OF THE HUMAN SPIRIT...

....AND...

## The FORCE That Lies

## Within ONE's Own FREE WILL!"

# The Mysterious Ancient Mayans

# P R E F A C E

**"All that we are is a**

**result of what we have thought."**

Buddha

563BCE – 483BCE

**L**ook Up...Look Out...Look Around...

at the Immaculate PERFECTLY-DESIGNED Universe all around you....AND NOW...LOOK AGAIN... with the knowledge and wisdom that YOU...and...Everything Around You...is intricately connected through a GRAND MATRIX which holds the entire Universe together and binds it as ONE Living Organism...

12:21:12:11:11:00:00                                        2012

This is the first book of a Trilogy on the multidimensional phenomenon of 2012. Throughout these pages you will find ONLY the clear, concise, scientific facts behind what mainstream, scientists world-wide have agreed will occur on December 21, 2012 at 11:11 AM.

To create a complete picture it has been necessary to draw from many sciences including physics, astrophysics, astronomy, archeology, anthropology, geology, biology, physics and quantum physics. Each fact is presented in an easy-to-understand way that will, hopefully, intrigue each and every reader to open their eyes to the greater landscape of this planet and it's environment...which we call home.

I have been extremely careful to present ONLY that scientific evidence that has been confirmed by at least three or more notable, highly-esteemed scientists or scientific sources. But~ it is important to note ~ that by the very nature of science itself all evidence, data and analysis (within any field of study) is in a continual state of transition. The reason for this is that as humans we are limited within our scope and ability to see the overall picture. There are a limitless amount of unseen forces that may affect our planet from beyond our solar system and our galaxy. So...science is and always will be, in a state of the changing.

I encourage all of my readers to research and reevaluate for themselves each and every area of interest, not only with respect to this book, but in every aspect of your life. This publication will provide you with a starting point from which to begin your own, personal JOURNEY towards 2012. As you research and reevaluate the phenomenon of 2012 be sure to keep an open mind. Even more importantly, always be open to LISTEN to your "inner voice"... your intuition ... as this will guide you to where you need to be.

## 2012 is a book of FACTS~Not FEAR

··· a word that is all too often associated with the time period of 2012. This book has been written with

the purpose of providing you with the Knowledge You Need to begin to understand the very important phenomenon of 2012 ~ a phenomenon that has never before happened in the history of mankind

on this planet. With this Knowledge will come

your own personal and unique Power....not only to understand, but to embrace and even become part of this precious time in Earth's history. Knowledge shall

lift you up, above the fear, and empower you to Flow through this heavenly, grand event, because in the end, the outcome to 2012 is – in my belief – as unique and as diversified as is each and every one of you!

The answer to 2012 may very well be different for each and every one of us on this planet we call home. Perhaps we are the creators of our own destiny...and the very fact that you have this book in hand proves that~

## There is a Unique & Personal Answer   awaiting Specially for YOU!

   Look up...look out...look around...look beyond...and listen to your inner self. Then you will find your personal answer...your own, unique destiny...perhaps a most glorious destiny that is awaiting you in 2012!

   Put on your seat belts, buckle up, and get ready for the ride of your life! No matter what you draw as your personal conclusion to the science I present  – one thing is for sure – you will never wake up and look at the world around you in quite the same way again! YOU have been chosen...and YOU are on your way to an exciting new world....a world that awaits you very, very soon...the new world of 2012!

### Hoping YOU Find YOUR OWN Unique & Special Answer...

### With God's Speed

### The Author

12:21:12:11:11:00:00                          2012

20

# Table of Contents
## Introduction

12:21:12:11:11:00:00                                                          2012

24

# PRELUDE

As we quickly approach our destiny with our long-fated 2012 Galactic Alignment scientists around the planet have noted that TIME appears to be LITERALLY "speeding up". Not only that but we have had GIANT LEAPS in our understanding of both our planet and the galaxy and universe which surrounds us in 2010 ALONE! Almost as if a veil is being lifted in "order that we may see." Yet, little ~ if nothing ~ ever reaches the main stream media and the public at large…and so much drama and confusion continues to surround this ever so critical event! As a result each and every one of us who is lucky enough to be even vaguely aware of this fast approaching NEXUS Event is left asking ourselves…"What is 2012 REALLY about? Is it merely the interpretation of ancient mythology, folklore and prophecy that has been  passed down throughout the ages much like some type of metaphysical Santa Clause, or is there something behind the date December 21st, 2012 that is truly tangible and real?"

What are the REAL scientific facts behind the 2012 phenomenon? Do current day scientists agree with the ancients on this precise date, and if so, what do they theorize will happen on December 21st of 2012? Is this something I need to be preparing for? What do I need to do in order to make sure that my loved ones and myself are safe during this very critical  time?

And lastly, what were the ancients trying to tell us, and how could they possibly have known so much about a date in time over 3,000 years into their future? Who were the ancient Mayans and is it by chance that their calendar system mysteriously ends on the date 12/21/2012?

All of these questions will be discussed and answered within **2012: A Trilogy**. I have chosen to separate each major topic into it's own book because of the depth and diversity of each subject, and so as not to bombard my readers with too much information all at once.

## PREPARE YOURSELF A SPECIAL PLACE

Before you begin reading this book...take the time to PREPARE YOURSELF A SPECIAL PLACE...Create a special place for yourself...your own private refuge from the world where you can escape...not only to read this book, but anytime you need to recharge yourself. If you don't have such a place...now is the time to create one for yourself. It should be your own unique version of a very peaceful, pleasant environment. Surround yourself with anything and everything that makes you feel relaxed, safe, and at peace.

When you are ready, get comfortable, light a candle and dim the lights...much like you the preparing yourself for a meditation.  Then clear your mind from all of the day's clutter ~ and try to approach each chapter with a clean, blank slate ~ void of any preconceived notions or beliefs.

## Become Apart of Your Special Place

Take a moment and try to picture yourself sitting on a very tranquil, tropical beach, enjoying the feel of the sun on your skin and the tranquil sounds of the ocean as it "plays" around you. The sand is a beautiful shade of white, and the water is very shallow and calm...a beautiful, clear Caribbean blue. Feeling tranquil and at one with your environment you decide to walk slowly to the shore and then wade slowly out into the warm, calm waters of the sea. Becoming ONE with the water and your environment time seems to "stop" or become nonexistent...you feel FREE, peaceful, and at one with God and his creations.

Then, from out of the calm sea, a very large, enormous wave emerges. As you watch it approach you know that you should be FEARFUL, you should swim to shore and RUN in order to escape it.

BUT, because you are feeling so "ONE" with the ocean, so peacefully a part of God and his creations, you feel no need to escape. Instead you greet the wave with openness, without any threads of fear. As it engulfs you, you do not fight it. Instead you willingly submit to it…allowing yourself to become a part of it.

Immediately you can feel yourself rising ~ as the wave lifts you up and you float gracefully on top of it on it's journey to the shore. When you meet the land the wave easily dissipates around you leaving you standing on the shoreline...amazed at yourself and amazed at the wave for not destroying you.


You realize now that had you run from the wave in fear it would have tossed and tumbled you into the sand, perhaps crushing you with it's weight. It could easily have devoured and destroyed you, yet it held you delicately in it's palms...and rather than destroying you it allowed you to become a graceful and beautiful part of it ~ almost as if it was protecting you ~ prizing you as a very special part of it's own existence.


As you take your place back on the beach you realize that perhaps this is a lesson that you can use everywhere within your life. Perhaps this is the way you should be approaching every moment, every event in your life. You realize that you need to become one with, and a part of, the flow of everything around you...and through that FLOW you can more easily, more successfully find your way...achieve your goals...all the while remaining in that peaceful and blissful state you discovered on the beach.


...This is the way of nature...

...This is the way of God !

12:21:12:11:11:00:00                                        2012

# I

# Crossing

a

# Chasm

of

# Millions of Years

# Crossing a Chasm of Millions of Years

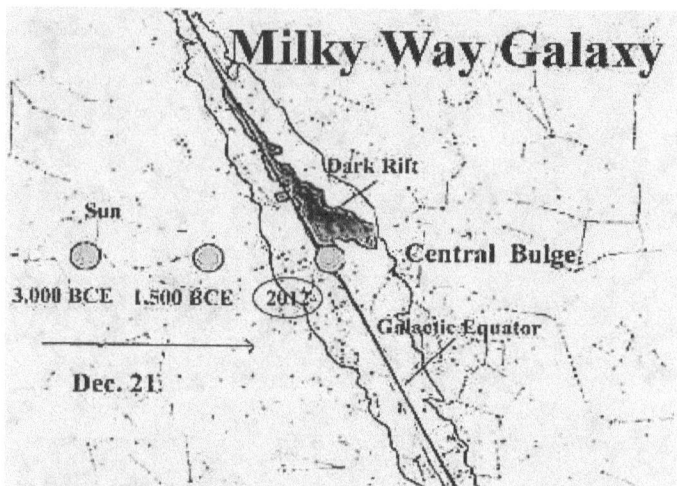

## The Galactic Center ~ The Dark Rift

# Crossing a Chasm
# of
# Millions of Years

D ecember 21st, 2012 ~ it is a PRECISE DATE
that has mysteriously been passed down throughout
the ages both within written texts and also within the
folklore and story-telling of almost all ancient cultures
everywhere ~ reaching into even the most remote
areas of our planet. It has even been referenced to
within many writings and passages in our present day
Bibles! How do we explain that ancient cultures
thousands and thousands of years ago could have
possibly known about this event? How is it possible
that civilizations so far apart in location and time, and
from areas all over the globe all point to the same date
thousands of years into the future? Only recently have
our present-day scientists been able to observe and
study these very elements of our galaxy...so how did
these ancient civilizations have the technology to
calculate these astronomical events so precisely? The
2012 phenomenon has a multitude of layers and
dimensions encompassing ancient prophesy, native
folklore, mystery, metaphysics, and...most importantly
...SCIENCE.

Yet, for some reason, the sheer DRAMA of the mystical and metaphysical sides of 2012 overshadow everything else.  So much so, in fact, that almost nothing has - to date - reached the public about the tangible, scientific facts behind the phenomenon....

...AND...

...that is the sole purpose of this book. Within these pages you will find the SIMPLE, SCIENTIFIC FACTS behind the 2012 event. As with the metaphysical MYSTERIES behind 2012, the sciences behind 2012 are also multidimensional in nature ~ drawing from the areas of   physics, astrophysics, geology, astronomy,  anthropology, archeology, biology, and even quantum physics.


Here are some of the most critical questions to be addressed within the following  pages:


# 1. ▪ What are the CONCRETE SCIENTIFIC FACTS behind the 2012 Nexus Event?


# 2. ▪ What do the worlds top, most highly-accredited scientists think will, or will not, happen on 12:21:12:11:11:00:00 at 11:11 AM?

**3.** What are the AGREED upon FACTS that
the scientific community is certain of ~ with
respect to the upcoming December 21st 2012
Nexus Alignment?

**4.** What is this "ALIGNMENT" that is spoken of
so frequently, and, why is it of such significance?

**5.** What are the plain and simple scientific facts
behind December 21st of 2012?

**6.** What do scientists foresee happening when
we cross this **Chasm of Millions of Years** ~ and
meet with Time Wave Zero at 12:21:12:11:11:00:00?

This book is dedicated to this ONE aspect of 2012
~ the concrete and provable SCIENCE BEHIND the
event. It incorporates all of the sciences necessary to
create the picture in it's entirety. Each science is
explained in easy-to-understand terms, and, MOST
IMPORTANTLY, all data and information contained
within these pages have been confirmed by three or
more highly-accredited, scientific sources which are
referenced within each chapter.

It is, however, critical to point out the fact that as important as the SCIENTIFIC side of 2012 is in order to understand and prepare for this unique time...the METAPHYSICAL aspects of 2012 remain equally important.

The metaphysics behind 2012, like the sciences, are also multidimensional in nature and exceptionally intriguing ~ and will be the subject of the following books in **"Trilogy 2012."**

## Galactic Center Region

## Crossing the Ecliptic

Our Sun will make it's crossing at a
60 Degree Angle to the Galactic Plane

# ON

# 12:21:12 at 11:11:00:00 AM

### Trilogy 2012/Book II :

**Books II and III will Focus on the following areas:**

**-A Day in the Life of a Geographical Pole Flip**

**- Safe Havens**

**- Survival Guide**

**- Preparedness**

**– Checklist for Survival.**

The metaphysics behind 2012 is most probably considered to be sheer MYTH by most who know about it...BUT it interesting to note that, perhaps NOT by COINCIDENCE, scientists and physicists from all around the planet have VERY RECENTLY come to a rare, unanimous consensus on the fact that metaphysics CAN and IS NOW scientifically PROVABLE, and easily incorporated into the standard laws of physics as we now know them!

Physicists are now able to prove even the most metaphysical phenomenon (psychic knowledge and even energy healing) through the study of subatomic and quantum physics...and experimentation with protons, neutrons and electrons on a subatomic level. This may not make the metaphysics more understandable to most people...but it does have highly SIGNIFICANT and GRAND implications for all of us...proving beyond a shadow of a doubt that we are all much, much more that the sum of our parts!

## We are ALL connected and interconnected
## both with our Planet and with
## the Universe as a whole...
### AND

along with this comes the realization that each and every one of us...like the universe itself...

goes far beyond the constraints of time & space!

Within each one of us God has created the basis for an eternal essence...that like energy and matter "can neither be  created nor destroyed." As written the texts and beliefs of almost every religion known to man...God has truly given each of us a piece of "*His Divinity.*" Yet, even the most devout believers have yet to fully understand, nor to be able to tap into,  this portion of who we truly are...and what we are destined to become.

Now, as we approach the calendar date of 2012, scientists and physicists alike are suddenly putting it all together. Almost as if time itself has lifted us up into it's arms, we are now, suddenly, able to link the mystery and the metaphysics with the basics of science, bridging together everything mankind has always known deep within his soul...mythology, ancient texts, folklore and modern day religion all coming together. Our instincts have always kept us very close to the truth, but it is only within the past few years that we can finally KNOW that our BELIEFS can be not only PROVEN but greatly and MAGNIFICENTLY ENHANCED....

# The Energy Field of

# The Great Matrix

## Artistic Rendering of

## Our Universe's Great Energy Matrix

# WELCOME
# ABOARD

## To Your Own
## Personal Journey
## to**12:21:12:11:11:00:00**...

...


**If you now have this book in hand**

## YOUR NAME is
## MOST CERTAINLY
**engraved upon on the**
## VIP Passenger List!

12:21:12:11:11:00:00                    2012

42

# II

Alignment
is
Coming
...But
Which One?

# 12:21:12:11:11:00:00

# 12:21:12:11:11:00:00
# Alignment is Coming
# But Which One

When I first heard about the 2012 PHENOMENON four years ago I became instantly captivated! I immediately set out on my very own MISSION to the STARS in order to discover and uncover the TRUTH about what was really going on. I have an education that includes the basics of physics, astrophysics, quantum physics and geology, and I have also been a metaphysical person since my first breath. Even so, I never in my wildest dreams envisioned the future writing of this book. But, FATE had different plans! Four years later, surrounded by stacks boxes overflowing with research and data from scientists all over the planet, this book decided to BE...began to evolve...and VERY QUICKLY took on a "life if it's own.

My very first MISSION was to establish the credibility of the date 12/21/2012. Was this just another rumor echoing throughout the blogoshpere based upon the ancient Mayan Calendar and their AMAZING calendar system. Or, did this date have a sound and provable astronomical and astrophysical credibility of it's own? If it did it would most certainly be noted and referenced to within our current-day scientific community?

I made it an absolute requirement to locate more than three, highly-accredited, scientific sources agreeing upon that fact, and I would only consider those scientific sources with highly credible, notable names...such as high-ranking universities, scientists and organizations including NASA and NRDO.

As soon as I began my research I was overwhelmed by shock and amazement! I found that ~ YES ~ there was a scientific community, world-wide, that was aware of and studying this upcoming phenomenon! And, not only did scientists around the world agree on the date December 21st, 2012, but NASA had also established a precise time for the "Galactic Alignment"...and that time was quite mysteriously also a binary number ~ 11:11:00:00 AM.

**There was ABSOLUTELY no controversy over the date or time of 12/12/2012...nor over the Galactic Alignment**...but there was some dispute over the physical and astrophysical effects the alignment may or may not have on our solar system and our planet Earth...a dispute which I quickly understood upon beginning my journey into the complexity of the event.

First of all, what precisely is this ALIGNMENT that has been spoken of and referenced to by ancient cultures all over the world going back in time to over 5,000 years? From a purely physical standpoint it is a

46

very unique alignment between the sun and the earth, both aligned together and sitting PRECISELY on the center-line or equator of our galaxy...which is called the Galactic Plane.

The calendar system of the ancient Mayans provide the best explanation for this event. The Mayans were completely obsessed with the sky and were in some ways the finest galactic time-keepers ever known to exist throughout the history of mankind. To present day, even with our sophisticated computer systems, satellites and telescopes, science has yet to put together an astronomical calendar that even comes close to that of the Mayans. Their calendar system surpasses all of the astronomical calendar calculations we have today, not only in the scope and depth of it's calculations, but also in it's absolute perfection and *on-target* accuracy pinpointing a precise date over 3,000 years into the future!

According to the ancient genius of the Mayans we reach this alignment with the galactic plane once every 26,000 years...crossing it over and over again as we bob up and down across the plane of our galaxy throughout the period of 250 million years, which is equal to one full Solar Galactic Year.

A Galactic Year is equal to the time it takes our Sun and her star system to complete one full orbit around the Milky Way Galaxy. Like our Earth year, taking 365 days in order to complete one full rotation in our orbit around the Sun, our Galactic Year is estimated to be between 200 and 250 million years long.  In other words our sun along with her solar system follows along an enormous, highly elliptical orbit throughout the Milky Way Galaxy, and it takes 250 million years to complete one total revolution. During one galactic year this particular earth/sun/galaxy alignment will occur over and over again...during it's 250 million year cycle...and culminating at 250 million years when we complete one full orbit throughout the Milky Way Galaxy, ending one Galactic Year and beginning a New Galactic Age.

# Our Sun's 250,000,000 Year Orbital Path throughout the Milky Way

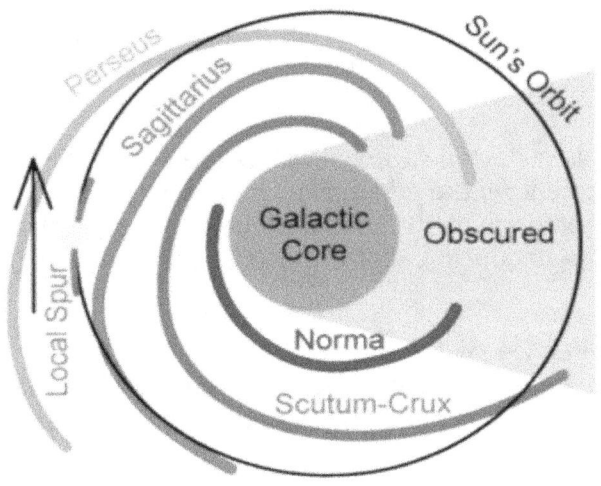

It is quite important to stop at this point and note that there are actually two separate and distinct types of alignment our solar system makes throughout it's million-year journey through time and space.

– One is an alignment with the center point or center plane of our galactic arm. Once again according to the Mayan's calculations we pass this center-plane every 26,000 years, taking 52,000 years to make one complete pass across our spiral arm.

– The second type of alignment is that with our solar system and the galactic orbit. This point is of some uncertainty, but some of the world's top astrophysicists believe that we bob up and down across our galactic path approximately 2.7 times throughout our (approximate) 250 million year cycle.

It is quite controversial with respect to which alignment we are addressing in the upcoming phenomenon of 2012. As we proceed, however, the picture begins to become much clearer.

When we take into account that our planet is estimated to be in the billions of years old, both of these alignments are obviously quite common events to Planet Earth and her solar system.  She has obviously survived both over many millenniums...so what exactly is so markedly different about this upcoming alignment in 2012? There are some who say that it is just another alignment and since the Mayans saw each alignment as the dawning of a New Age it was a notable point in which to end their long and elaborately detailed calendar. They had to choose an ending date somewhere, and they were already over 3,000 years into the future...Super Case in Point ...CASE CLOSED!

It makes perfect sense unless you begin to dig a bit deeper into both the Mayan civilization and the astrophysical side of this event.

First of all there is the astonishingly, amazing fact that the Mayans BEGAN their calendar not at their present day...but instead they moved forward to the alignment date of 12/21/2012...marking this as their end date, and the ending of the current Grand Cycle. They then worked their way backwards in time to their current place within time and space 3,000 years minus that date, at which time they could place themselves on the correct galactic point within the current solar/galactic orbital year.

However, it is also very important to note that this particular galactic alignment was not only just another galactic milestone...but also a very **SACRED EVENT** to the Mayans. It was to be a time of enlightenment, the **Dawning of a New Age**, and a time during which they would be very close to their **God-Force.**

And, MYSTERIOUSLY, they have made sure that this information was never lost, and would be passed down through the eons by way of their numerous intricate drawings and carvings...all created to depict this very event. Most all of their drawings make reference to the Dark Rift which they called "Hunab Ku". This is the very dark and cloudy area at the precise center of our Milky Way Galaxy.

# .....Then......
# without warning..
# the entire Mayan Civilization just
# VANISHED......
# SIMPLY DISAPPEARED
# off the face of the Earth!

And, in doing so, they left EVERYTHING behind! It was as if they just got up one morning and walked off...without so much as an extra piece of clothing or a small supply of food. There is no evidence of warfare or invasion from a nearby tribe. And, there is no geological evidence of any major climactic event having taken place at that time. Yet, they fled so suddenly that it appears to have been some sort of very critical reason which they had no previous knowledge of. Everything was left behind in perfect order...except for the Mayans...whom have never reappeared anywhere on the planet again!

# Mayan Artifact Depicting
# Solar Cycles & the Dark Rift

And, leaving us with yet another mystery to add to the 2012 saga! ...Back to the information they laboriously left for us...with a bit more thought one can see that even if such an alignment would occur a thousand times per galactic year, each alignment would take place at a different point within the sun's orbit. Much like our planet's Spring, Summer, Fall and Winter our sun moves along an elliptical orbit during which time it is closest to the galactic center, and then progressively further and further away until it winds back towards the Galactic Center and ends another 250 MILLION YEAR ORBIT around the Milky Way Star System.

The Mayans depicted the alignment of 2012 as bringing us in sync with our Galactic Center, while also making it very clear that we will be about to END one AGE and BEGIN a BRAND NEW AGE. We must also consider the fact all of the texts and symbols the Mayans used to describe this event depict it as a HIGHLY SACRED EVENT, and one that brings them in touch with both eternity, their "God" force, and a planet by the name of MAYA. Maya is a planet in the Pleides Star System, and the ancient Mayans claimed that this was their original home.

After taking into consideration all of the above facts, we must at least consider the fact that this particular alignment may not be just one amongst the thousands...this alignment **may PERHAPS be the Grand Finale Alignment of our current 250 million year cycle!** That is the only one of the alignments that would bring us directly in contact with their Hunab Ku or our Galactic Center.

Suddenly, it all begins to make sense....why an ancient race would take the time to painstakingly work forward in time thousands of years in order to begin their calendar system....and as to why this date in time believed to be of such EXTREME IMPORTANCE to them that it was depicted in drawings and carvings over and over again.

# And so...The Mystery Opens...

## ...Establishing the fact that...

# WE WILL REACH CRITICAL ALIGNMENT on the 21ST DAY OF DECEMBER 2012 AT 11:11:00:00 AM!

What still needs to be established is which one of the numerous galactic alignments will we be crossing on that date? The Galactic Orbital systems for all galaxies within our universe are precisely the same ~ all follow the exact same overall design ~ or what astrophysicists call an Intelligent Design. This is GREAT news because it is, perhaps, the closest science will ever get to admitting the fact that GOD is a basic factor in all that was ever created within our universe. As a result of its' Intelligent Design, all galaxies and all of the star systems within them follow precisely the same laws. So...like an enormous grandfather clock each tiny cog relates to the next cog above it...and this goes on throughout all of the many cycles until we once again arrive back at the Galactic Center. At this point all of the cogs will be lined up to Time Wave Zero ~ 00:00:00:00 ~ and all sub-cycles will simultaneously close at precisely the same exact time.

Taking this into account, we can see by the data that we will most certainly be closing a 26,000 year cycle at this time...but it may not end at that cycle...we just MAY also be closing out another 250 million year Galactic Orbit. This would explain the high **SACREDNESS** of this event to the ancient Mayans, and the fact that they believed this alignment would bring them into contact with their God Force and then mark the beginning of a brand NEW AGE.

Could it be possible that astrophysicists are aware of this fact?  Perhaps they choose to leave it out of their theories in order to maintain peace and tranquility amongst the people.  In the end, panic and fear are extremely negative forces,  and the scientists most probably believe that  we would have a much greater chance of survival without them.

In addition,  all scientific records for our planet indicate that it has now been 250,000,000 years since our  last MAJOR EXTINCTION, and 65,000,000 years since our last MINOR EXTINCTION!  So, we are obviously VERY CLOSE to the end of our current Galactic Orbit.  The question remains...just HOW CLOSE?

In either case,  what do the top scientific minds from all over the planet theorize **MAY or MAY NOT** take place on our planet at the time of this critical galactic crossing?

# II

# Journey
# to the
# Dark Rift

# ...A Solar Year

# Our Slow Journey to

# The Dark Rift

A = 8000 BC

B = 4000 BC

C = 2012 AD

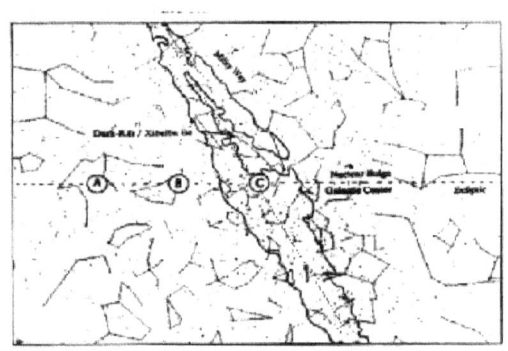

图1. 冬至的太阳和银河系中心重合 (C: 2012年12月21日)
A：8000 年前，B：4000 年前。
红点斜线是银河系赤道，
黑点水平线是银河系黄道。

# Journey to the Dark Rift...

# A Solar Year

We shall proceed into the science behind the 2012 Galactic Alignment. Before we do ~ it is important to set the Galactic Stage ~ with just a brief overview of our Milky Way Galaxy. Quite briefly, here are the facts as science currently defines them:

Our Milky Way galaxy has a total of six spiral arms. From the Galactic Center moving outward these arms are identified as: The Norma Arm, The Centaurus Arm (Scuutum-Crux Arm), the Sagittarius -Carina Arm, The Orion Arm (part of a small Local Spur), the Perseus Arm and the Cygnus Arm.

———

The Milky Way Galaxy is home to about 200 billion stars.

———

Our Sun and her solar system are currently located on the innermost rim of the Orion Arm.

Scientists calculate us to be 8000 parsecs from the Galactic Center.

Our Milky Way Galaxy ~ along with 50 of our closest neighboring galaxies ~ make up what is called The Local Group, and our Local Group is part of the Virgo Cluster.

It is also quite amazing to note that our Galaxy is actually part of a Twin Galaxy. We are part of a binary galaxy system with another galaxy called The Andromeda Galaxy. Both galaxies are merging towards each other and will, at some point in the very far future billions and billions of years from now, merge together to become one galaxy. Or, they will end up destroying each other in the process.

Our Sun and her entourage of planets travel along a highly elliptical orbit throughout the Milky Way Galaxy. According to most scientific estimates it

―

takes us somewhere between 200 million and 250 million years to complete one full rotation throughout our orbit. This is called our Galactic Year.

―

During it's long, laborious travels our sun moves throughout the Milky Way in the direction of the star Vega and the constellation Hercules, at an angle of approximately 60 degrees towards the galactic center.

IF DECEMBER 21ST, 2012 IS CORRECT AS BEING THE LAST DAY OF OUR SUN'S 250 MILLION YEAR ORBIT our SUN and all of her planets, including our precious 3rd rock from the Sun, will be crossing the Galactic Equator and beginning the passage through the Galactic Center (Dark Rift) on this precise date. So, let's take a quick look at the overall neighborhood of our Galactic Center.

# The DARK RIFT

At the center of our Galaxy ~ and as observed at the center of all galaxies we are able to observe at this time ~ is a large, elongated area of dark, dense matter often referred to as a Black Hole. Black Holes are not completely understood by science at this time, but it appears to be obvious that these Black Holes

61

function much like giant super-engines and are critical to the formation and functioning of every galaxy within our universe.

If you were able to observe the skies on a very clear night you may be able to see the Galactic Center or Dark Rift with your naked eye. It appears as an elongated, cloudy area extending horizontally in the night sky.

The Dark Rift is a highly complex region that is ALWAYS obscured by a series of dark, dust extending from the constellation of Sagittarius to Cygnus. In addition to the continual turmoil of dust, the Dark Rift is also filled with tons of space debris including asteroids, meteors, comets, super nova remnants, and an enormously, giant *star* called Sagittarius A ~ which is estimated to be approximately 1000 times larger than our sun. Scientists believe that Sagittarius A may not actually be a star at all, but may be the actual black hole at the center point of our galaxy. Very recent discoveries have uncovered yet another mystery...in that Sagittarius A is also the source of intense, radio-like emissions. In addition to the visible inhabitants of this zone it is also overrun with an unknown number of extremely strong magnetic fields and currents...some of which we are not even able to observe nor calculate.

# WHICH LEADS us to SOME VERY IMPORTANT
## QUESTIONS.....

Why is crossing the Galactic Center so important? Isn't it just the beginning of another 26,000 year cycle? Or is it possible that, in addition, it is also MUCH, MUCH MORE?

## 12:21:12:11:11:00:00

Long Count: 12.19.19.17.18
Days since Creation: 1871998

Calendar Round: 2 Etz'nab 1 K'ank'in
Lord of the Night: G7
Katun (end date): 4 Ahow 3 K'ank'in

Gregorian Date: 21 December 2012
Julian Date: 8 December 2012

Correlation Constant: 584285
Julian Day Number: 2456283

### Mayan Long Count Calendar

**Consider the possibility that this critical date could be both the end of another 26,000 year cycle...**

## AND ALSO

## The FINAL DAY of our Sun's
## 250 MILLION Year Cycle!

In Earth terms this is equivalent to one full orbit of 365 days, and one complete Earth year. In the Sun's case, one full orbit is around 250,000,000 years, which is equivalent to one full Solar Year. This 2$^{nd}$ alignment would also need to include the 26,000 year alignment and the close of ALL CYCLES ~ since all cycles must reach 00:00:00:00 alignment at precisely the same time. After comparing this conclusion to the Mayan Calendar we find that it fits together with it much more exactly. AND ~ it also matches all of our planet's current paleontological records ~ along with the changes we are now seeing both on our planet and throughout our entire Solar System.

In either case, it sounds like one SUPER New Year's Party to me! BUT, do we even know where are we in our galactic orbit today, in 2010? Or is science not able to agree on our present orbital location? And, if we indeed are fated to cross this sacred galactic plane in 2012...what do top scientific minds around the world theorize could or could not happen on the surface of this planet we call home? These are just some of the great questions which will be answered and explained within the following pages. As we proceed we will "set the stage" for each topic in order to provided a brief background and a basis for putting the pieces of this enormous MYSTERY together.

# IV

## The
## Horizon
## Is
## Upon Us

# The Horizon
# is Upon Us

W e have created a wonderful civilization filled with technologies we never even dreamed of...a world with no limits...in which we believe that anything is possible! We have conquered each and every obstacle that has stood in our way!  BUT, as we race through each hectic day, we have little or no time to STOP and Take a Look Around Us...

I ask each and every one of you to STOP, just for one moment, and consider this...Should we be bracing for tomorrow?  Are we truly living "on the edge of time" as foretold by countless ancient texts and prophets from around the world for hundreds and hundreds of years?

# STOP HERE...for just one MOMENT...

Open your mind and just allow yourself to PLAY. Evaluate, scrutinize, and verify all of the new information that has come available to us over the past several years...all of which is presented within this book.  It is so important, in fact, that it MAY have a direct impact upon the very immediate future of both yourself and your loved ones.

It has been wisely said that "those that do not learn from the past are bound to repeat it." . And, we all know the saying...KNOWLEDGE is POWER! Consider for one moment that KNOWLEDGE is not only POWER...but the KEY to YOUR very survival...the KEY to the Golden Door that will open for  YOU, allowing YOU to enter and cross over the EDGE of TIME and into the new world that will follow.

When all of the greatest of the scientific minds are put together....and all of the information from their various areas of study are pieced together...an entirely NEW picture of our planet's history emerges! And it is a picture which has not ever been presented to the public!

Perhaps this is because it has been deeply imbedded into our human nature to reject any information which we feel does not fit with our version of reality as we know it to be.  In fact, it is almost impossible for us to imagine our world any way except the way that it currently is. We are imbedded into the here and now...and to us that is a constant. It is almost impossible to even consider our planet ever changing in any way.

**The Horizon Project** is a group of the best scientific minds on this planet...and their only mission is to piece together such a PICTURE. The Horizon Project is composed of an elite team of scientists, comprised of a group of the most brilliant scientific minds and researchers from around the world in every scientific specialty. They formed together with the unique mission to accumulate and analyze ALL of the evidence, from every scientific specialty, in order to create a complete picture of our planet's history... and future.

"Should we be bracing for tomorrow, as foretold by countless ancient civilizations, texts and prophets for thousands of years?"

"Are we truly living at the edge of time?

The first puzzle the Horizon Project sought to decode is that of GAPS...gaps in human history...gaps that have gone unrecognized by mainstream scientists, but gaps that have been so critical as to have reset human evolution many times throughout the history of our planet.

Interest in these gaps stemmed from newly discovered evidence recently uncovered by archeologists that paints a picture containing some extremely startling facts. It is a commonly held belief that our current civilization is at the pinnacle of advancements and technologies never before known to exist on this planet.

However, recent advancements in research have found that ancient civilizations have made use of some very highly evolved technologies...paralleling and even matching those that we use today! In fact recent findings show the use of electricity and sophisticated technical advancements going back thousands and thousands of years! Even more startling, researchers have found evidence of successful brain surgery being performed 7,000 to 9,000 year ago, along with other ancient technologies such as the making of pure forms of glass and steel, an ancient 2,200 year old battery that had been discovered and then overlooked back in 1938; along with ancient calculators, evidence of the use of electricity, and even remarkably complex, detailed and accurate world maps. The Smithsonian Institute states that steel was being made in furnaces over 7,000 years ago, and drawings found in Egyptian

Pyramids depict bulb-like components that have a startling resemblance to the modern day light bulb.

So...if, in fact, the Earth has a history of highly technological civilizations...what has happened to them? Where are these GRANDE CITIES from the ancient past?

Within the past twenty years we have finally developed the technology required to study the depths of the ocean floors...and what scientists have found are cities...many, many cities...cities and civilizations of which we have absolutely no record! These massive, ancient cities have been located off the coasts of Cuba, India, Australia and Japan...and more are being discovered every year. Some have sunken to depths of almost 2,000 feet beneath the ocean's surface and all are mostly intact, as if they simply sank into the ocean. Not just the city itself but the entire land mass upon which the city was located.

As startling as this newly uncovered information may be...the most critical question for us to decode is the cause or event that destroyed these ancient civilizations? To answer this question the New Horizon scientists studied two very critical areas of data, both revealing untainted evidence of our planet's ancient past.

The first of these is that of the ICE CORE SAMPLES taken from the arctic and Antarctic regions. Polar ice core samples contain layer upon layer of ice, each LAYER containing untainted particles and evidence showing atmospheric conditions that have been preserved and unchanged for thousands and millions of years.

Geological studies of these samples have revealed some extremely startling evidence.  For some mysterious reason scientists have found layers of cosmic dust sealed into the ice, repeatedly and in a cyclical pattern. This led them to conclude that whatever is causing this cosmic dust to fall to the ground happens on a regular basis...and, it is not a long-term event, but rather an event that takes place within a very short period of time...a period of merely days and weeks, or perhaps months!

The second most critical study comes from the middle of the Atlantic Ocean. This area is known as the Mid-Atlantic Ridge and consists of an enormous volcanic mountain range running vertically up and down the middle of the Atlantic Ocean.

As with the ice core samples, the LAVA CORE SAMPLES from this region also reveal information about the climactic conditions present on our planet at the time the lava solidified.  Due to the time it takes the lava to solidify, the current magnetic field of the planet is literally recorded into the  strata structure of the lava core samples...providing scientists with a precise history of the magnetic field of our planet going back thousands and thousands of years. Again, these lava core studies show extremely abrupt changes taking place within the Earth's magnetic field...and, again, such changes are both cyclical and taking place in a reoccurring pattern.

## THE MILLION DOLLAR QUESTION

The MAIN question this scientific group of experts needed to answer is quite obvious:  What type of event could cause such extreme, immediate planetary changes, and not only that...but what type of event is capable of creating these abruptly devastating changes to occur at regular, cyclical patterns in Earth's history?

## AND...THEY CAME UP WITH A UNANIMOUS CONCLUSION!

There is only ONE plausible phenomenon that would explain these extreme anomalies, and only ONE EVENT that is capable of causing such ABRUPT changes to our ENTIRE PLANET, while also proven to be CYCLICAL and REOCCURRING over geological time?  That event is referred to as a **geographical pole shift!**  And it is a phenomenon that mainstream scientists agree to be quite real!  In fact, any slight disruption from an outside force would be enough to throw our planet off it's current axis...much like a top spinning on a table.  It is also important to note at this point that our planet is not only tilted on it's axis...but it wobbles as well!  So...exactly how STABLE is our planet?

The idea that we live on stable, solid ground is merely an illusion.  In reality, the surface layer of our planet is quite thin and very fragile.  Our planet is made up of the core, the mantel, and the surface or crust.  During a geographical pole shift the core and the mantel actually shift counter-clockwise to the thin surface crust....creating a force so extreme that it literally changes and rearranges the surface structure of the Earth's surface instantaneously!

To our planet as a whole it poses no threat.  The Earth simply flips ~ making a extremely fast 180 ~ and then just as quickly reorients and stabilizes itself in space and attains it's new axis and orientation with the sun.

AND, life goes on.  However, to many of the inhabitants on the surface of the planet these abrupt and extreme changes can be devastating!  This is due to the fact that the surface layer and everything living upon it goes through a **major, traumatic, and IMMEDIATE transformation!**

Mainstream scientists currently believe that our planet has endured at least three of these geographical pole shifts within the past 10,000 years alone...a time frame that in geological Earth years makes these events as common as the changing of the seasons!

For example, new advancements in technology have enabled scientists to discover huge petrified forests underneath the ice in the Antarctica. These forests contain trees as high as 80 feet tall and having foliage indicating those of warm climate trees ~ proving that this area was once warm and tropical and swarming with life. Even more startling was the discovery of ancient mammoths that had literally been flash frozen in time while enjoying a mouthful of lush foliage which remained still intact and preserved within their jaws. The event took place so suddenly as to not even enable these creatures to complete their final morsel of food!

During this time period our planet was sitting completely straight and perpendicular to the sun, creating a planetary-wide environment known as "perpetuum vernnum" or the "eternal spring." Today it is well-known that our planet is slightly tilted on it's axis, and is warped (or bulging) along the entire circumference of the equator. And, not only is our axis tilted and warped, but our planet wobbles as well... leaving us to ask the question, "In reality, how stable is our Earth?"

The Earth is comprised of three distinct layers: the core, the mantle and the crust. The crust, upon which all known life resides, is actually a very thin layer comprised of loose fitting material capable of great flexibility and change. In addition, it is broken into many moving parts called tectonic plates and it is upon these plates that all of the continents and oceans reside. As our planet experiences minor changes these plates move and slide against each other causing earthquakes, volcanic eruptions and tsunamis. Finally, a geographical pole shift, when caused by the JOLT of passing through the center of our galactic plane, is the ONLY EVENT scientists know of that could cause the layers of cosmic dust as found within the polar ice core samples. This is due to the fact that our galactic center is literally littered with cosmic dust and debris that is sucked into our galaxy from the outside environment. Our galactic center or black hole acts as a major vortex pulling and sucking everything into it's already enormous mass. And, this event of passing through the Dark Rift area of our galaxy is also one which occurs on a regular, cyclical pattern. Hence, we see the ice core samples layered with cosmic dust on regular and reoccurring intervals.

To obtain a more scientific look at this entire picture one scientist stands out above the rest in terms of his research and expertise. That scientists is the world-renowned physicist Dr. Brooks Agnew....whose research is more than deserving of it's own chapter...which follows.

# V

# Sciences
# Best Kept
# Secret

12:21:12:11:11:00:00                                        2012

78

# Sciences Best Kept Secret

Perhaps the best kept secret within the scientific world today is Dr. Brooks Agnew! This book would not be complete without his *take* on the entire subject of 21 December 2012...and what he believes may or may not occur at that critical time.

Dr. Agnew is a world-renowned physicist with over 30 years of ongoing scientific research in the fields of physics, astrophysics, geology, archeology and more. Dr. Agnew has published thousands of technical papers in the fields of advanced mathematics and physics, and is also the author of two best selling books. He has also made invaluable, ongoing contributions to organizations such as NASA and the Jet Propulsion Laboratories; and, in addition to his current work is also the host of one of the world's most popular scientific talk radio shows on X-Squared satellite radio, every Sunday from 8PM to 11PM EST.

Dr. Agnew confirmed the fact that we are about to complete a 26,000 year Precessional Age, and will be entering into the Age of Aquarius with the crossing of our galaxy's center plane on December 21st, 2012 at 11:11AM.

He added that this event and date are "well marked on all kinds of calendars, the most famous being the ancient Mayan Calendar." He adds that it has also been a part of the information handed down to us by many ancient cultures via ancient texts and stories.

When asked about his opinion regarding geographical pole shifts, Dr. Agnew stated that "geological evidence points to the fact that our planet has undergone one or more pole shifts in the past" and that, in his opinion, "this is a repeating, reoccurring, cyclical event for our planet." He evidences the strata structure of the lava rock in the arctic circle, which he states "shows a broad band of magnetic alignments" indicating that the Earth has held many different alignments with the sun.

Dr. Agnew describes a geographical pole shift stating that the crust of the planet actually counter rotates against the core, literally spinning over like a top on a table, and then quickly reorienting itself in space with a new geographical alignment. Therefore, the sun may come up in the east or the north, and the entire surface of the globe is affected along with weather patterns, migratory patterns, land masses, oceans and bodies of water, etc.

Dr. Agnew also states that at one time our planet had been in perfect alignment with the sun, as evidenced by geological studies of the arctic regions of our planet. Studies show that these regions were once warm and swarming with life.  Then something happened, "probably cosmological," that was very catastrophic to the Earth and led to a pole shift ending with our new position in space which is both tilted and leaning towards the sun.

He added that theses events "as seen in space" are observed to happen "quite suddenly."  Most scientists such as himself believe that we will have precursor events leading up to a geographical pole shift...such as an increase in the frequency and intensity of storms, hurricanes, volcanic eruptions, earthquakes, tsunamis, etc. The pole shift itself may occur one or more times and can be devastating to the surface of the planet due to land mass shifts and major tidal waves and flooding. Dr. Agnew estimates that a pole shift would  "stack the ocean up to between 5,000 and 6,000 feet on one side of the planet." Many physicists also state that a pole shift of this type would be accompanied by extremely high winds up to 700mph. These winds will be the strongest at the equator plus and minus 35 degrees both North and South.

Where rims and tectonic plates come together the land could split apart, and parts of land masses in these areas could simply sink hundreds to thousands of feet under the ground.  Islands and pieces of continents could also split off and simply sink into the depths of the ocean.

In fact, Dr. Agnew points out that new technology has enabled scientists to now delve deep into the depths of the oceans and there is evidence that entire continents have broken off and sunk to up to 2,000 feet beneath the ocean's surface.  Evidence of major cities from past civilizations have been found off the coasts of Cuba, India, Australia and Japan...and more are being discovered with each passing year.  These are ancient underwater cities, massive cities from previous civilizations of which we have no record. Scientists have also discovered huge petrified forests under the ice of Antarctica...along with the famous mammoth found literally flash frozen in time with a mouth full of exotic, tropical  foliage still in his mouth!

...HOWEVER...Dr. Agnew feels it important to note that this event is of no consequence to our planet as a whole.  It has experienced these shifts on a regular basis for millions and perhaps billions of years.  To our planet it is normal phenomenon just as is day and night, spring, summer, fall and winter. Our Earth very quickly reorients itself in space...

# ...AND LIFE GOES ON!

Dr. Agnew goes on to add that this event, the end of this Precessional Age, does not pick on our planet alone. Each and every planet in our solar system, including the Sun herself, are currently experiencing the precursor affects of this alignment. This subject will be discussed  in the following chapters.


**Final Quotes from "Bracing for Tomorrow" interview with Dr. Brooks Agnew, hosted by Senior Project Researcher Brent Miller:**


**Brent Miller:**  "...So, Dr. Agnew, you believe we will see clear signs that we are approaching this major intersection in space and such signs would be a precursor to the pole shift event?"

**Dr. Agnew:**    "Yes.  The signs are already here."


**Brent Miller:** "Dr. Agnew, could a geographical pole shift explain all of the geological evidence and effects we have seen in the past?"

**Dr. Agnew:** "Yes."


**Brent Miller:** "Do you believe this type of event could also be the cause of  the reoccurring layers of cosmic dust as evidenced within the Antarctic ice core samples?"

**Dr. Agnew:** "Yes."

**Brent Miller:** "Do you believe this catastrophic event will happen again in the future?"

**Dr. Agnew:** "Yes."

The Horizon Project's DVD, **Bracing for Tomorrow,** can be found at

**www.TheHorizonProject.com.**

It is a riveting and spell-binding, scientifically-based presentation of the 2012 phenomenon! If you choose to see only one DVD on this critical subject...this is THE ONE!

# VI

# Are We Living
# at the
# Edge
# of
# Time?

# 12:21:12:11:11:00:00

# Are We Living at the Edge of Time?

Are we literally living at the beginning of the edge of time? From ice and lava core studies (as discussed in Chapter 3: Paleontological Evidence) scientists now believe that our planet has seen at least three geographical pole shifts within the last 10,000 years. This is an extremely short time frame within geological years, and taking into account that our planet is over one billion years old.

The good news is that these abrupt shifts are a common, reoccurring part of the Earth's climate...much the same as the changing of our seasons from winter to spring, and summer to fall. A Pole Shift is also as common as the rotation of our planet bringing us 12 hours of daylight and 12 hours of darkness. The GOOD NEWS is that it IS POSSIBLE for some of us to survive such enormous shifts and upheavals. In fact, as with all of nature, such changes are most likely a form of the renewal or refreshing of life on our planet.

Fossil records indicate that the most recent pole shift took place approximately 13,500 years ago. According to world-renown physicist, Dr. Brooks Agnew, it takes only one catastrophic event for our planet to spin into a geographical pole shift.

However, the Earth recovers itself within a very short period of time...quickly reorienting and stabilizing itself in it's new position within it's orbit.

Dr. Agnew points out that it is man's civilizations that are devastated by such earth shifts due to the fact that a polar flip very quickly affects the entire surface of the planet. He point out that we live on a very thin crust we call the Earth's surface which is comprised of many tectonic plates. The centripetal force of these abrupt pole shifts forces our planet to quickly flip or shift in space...pushing the crust of the Earth's surface upwards and out. This immediately shifts the positions of all continents and oceans, and can be extremely devastating to all life on the planet.

The global havoc created by a geographical pole shift includes immediate eruption of many active and inactive volcanoes, super storms, fire, flooding , and ~ most alarmingly ~ tsunamis. Dr. Agnew theorizes that a pole shift will most likely, and quite easily, generate a major tsunami, stacking the ocean 5,000 to 6,000 feet high on one side of the planet!

A second major problem, according to Dr. Agnew, is that of wind. He theorizes that such a shift would create winds up to 700 mph everywhere on the planet for a minimum 72 hours, and most likely lasting one to three months. The effects of these wind storms will be the worst from a latitude of 35 degrees N to 35 degrees S. They are the causal effect of the fact that as we "flip" in space the inner core of our planet counter rotates against the crust, creating an enormous magnetosphere.

He predicts that with current population such an event would probably extinct about 95% of life on planet Earth. The most vulnerable places being near to any body of water, and near to any tectonic plate conjunction. Dr. Brooks Agnew currently hosts one of the most popular science shows on talk radio.

## Netale Chart:    21 December 2012

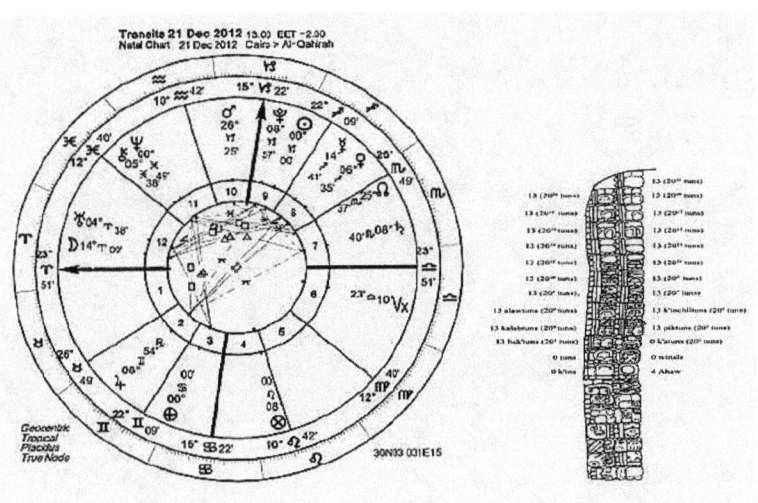

## Transite:    21 December 2012.

12:21:12:11:11:00:00                         2012

# VII

## 12:21:12:11:11:00:00
## Time
## Wave
## Zero

## Visual Depiction of the Hopi Prophecy

The Hopi's ancient ancestors predicted the above Earth changes to occur after the appearance of the "Blue Star." According to their legend, the ocean will rise to the "third mesa"...bringing the west coast of the United States to the approximate latitude of the Rocky Mountains.

# 12:21:12:11:11:00:00
# Time Wave Zero

The Galactic Center or Dark Rift can often be observed in the night sky with the naked eye as an elongated dark and cloudy area. The region is quite thin and elongated with a spherical shape. According to all known scientific sources our solar system is scheduled to cross the Galactic Equator at precisely 11:11AM on the 21st day of December 2012. Scientists estimate that it will take our planet 60 to 70 hours to complete our passage through the Galactic Plane. This means that we shall begin our passage on December 20th ~ be at Time Wave Zero at 11:11AM on December 21st ~ and complete our passage through the Galactic Equatorial Region on December 22nd of 2012.

Current day physicists and the ancient Mayans both agree that our solar system and her entourage of stars bob "up and down" across this plane during our 250 million year journey around the Milky Way Star System. If we stop at this point for just one moment and look back over the paleontological history of our planet, we see that the most major and devastating Earth changes and extinctions took place every 250 million years, with secondary extinctions occurring approximately every 65 million years....or at each ¼ point within our Sun's galactic orbit. The last major extinction has been dated at approximately 65 million years ago. According to First Science (www.firstscience.com), "Today, it has been 65 million years since the last major extinction event on our planet."

# The Dark Rift

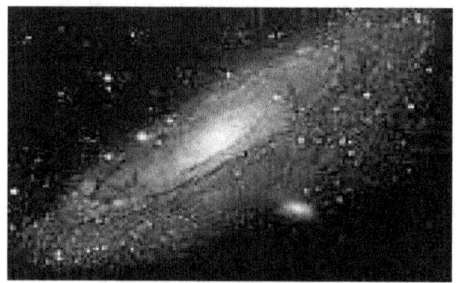

## is The Galactic Center

It has also been around 250 million years since the Great Permian Extinction, which most obviously marked the end of our last Solar Year, and the beginning of our current Solar Year. The question we must all ask ourselves is this: As we complete the current 26,000 year Precessional Age, is it possible that we are also completing our 250 million year Galactic Orbit? Could this be what the ancient Mayans were attempting to convey through their very detailed and elaborate drawings and carvings depicting the Dark Rift (which they called "Hunab Ku") and the Galactic Center? Could this be the reason they chose to go forward in time 3,000 years to the end date of December 21st, 2012 to begin their calendar system...and then work backwards to the planetary alignments as they were at their current point in time? Thereby they were able to place themselves at the correct point within our galactic orbit, and this was an obvious beginning from which to map out all orbital sub-cycles? Yes...as hard as it is to believe THEY WERE THIS ADVANCED...AND PRECISELY ACCURATE in their time-keeping!

To the ancient Maya 12.21.12 was also much, much more than just the birth of a new Solar Year, this particular passage was seen as an event of such significance that it was considered to be SACRED! It marked the BIRTH of BOTH a new Great Cycle and an **entirely new** Grande Precessional Age ~ again pointing to the possibility that we are also about to complete an entire Solar Year...or a 250 million year Galactic Orbit.

The ancient Mayan calendar began on August 11, 3114 BC and ends 5,125.36 years later on the Mayan date of 13.000.000...which equates to our calendar date December 21, 2012 AD. Within their elaborate artwork which accompanies their calendar system they depict the Galactic Center with what they called **"The Mayan Sacred Tree."** This is a visual depiction of the precise point of our "Celestial Crossroad" or the crossing point Equator (ecliptic). In addition, as our Sun makes it's orbit throughout the Milky Way

Galaxy it is important to take note of it's "apex" or the direction and angle in which it travels. Currently the Sun and the Earth are traveling at an angle of approximately 60 degrees towards the Galactic Center, and towards an enormous star named Vega (or Sagittarius A) near the Constellation Hercules.

Scientists, through the technology of the continent-wide VLBA (Very Long Baseline Radio telescope), were able to accurately observe a radio-wave emitting object named Sagittarius A that has been thought to be at the exact center of our galaxy. According to the National Radio Astronomy Observatory, "The measurements we made with the VLBA place Sagittarius A very close to, and most likely at, the exact (dynamic) center of our Galaxy. The new data also indicates that the minimum mass for this object is about 1,000 times the mass of the Sun....and strengthens the idea that this object contains a black hole about 2.6 million times more massive than the Sun."

The Milky Way's center is a very complex region containing not only Sagittarius A but also numerous supernova remnants and magnetic features. It is always obscured from optical telescopes due to the enormous amount of dust within the region. However busy our Galactic Center may be on any give day...it is going to become much busier on the 21st of December 2012...as it appears that many of the celestial bodies plan to make an appearance for this very special occasion!

# Rare Celestial Events in 2012

A very RARE transit of Venus will take place on June 5th & 6th of 2012. Venus will also conjunct with Alcyone~the central star of the Pleiades Star System. It is interesting to note that there is a planet named **Maya** within the Pleiades Star System...and that the ancient Mayans believed that they originated from within the Pleiades Star System.

–       Two MAJOR solar eclipses will take place in 2012.

–       On May 20th, 2012 our Sun and moon also conjunct with the Pleiades Star System.

–       On November 12th, 2012 our sun and moon conjunct with the Constellation Serpens ~ known for representing eternity.

–       Three major planets ~ Pluto, Neptune & Saturn ~ take position to form what the Mayans refer to as the **"Finger of God"**.

–       During this time we come very close to a star named **GM.**

–       We also come very close to the Trifled  MZ0 Nebulae.

In addition to ALL of the above,  the Galactic Center ~ as stated earlier ~ is a very complex and busy location in and of itself.  In addition to the massive amount of Cosmic Dust always present within this region, there is also a mixture of space debris that has been sucked in to our galaxy by the enormous vortex-like pull of it's giant Black Hole.

# Mayan Depiction of the

# "Finger of God" ~ Eternity

## ~ HUNAB KU (Lt) & Our Galaxy (Rt)

This debris includes small and large meteors, floating rocks and other mixed cosmic debris...again relating back to why it has been named The Dark Rift by the ancient Mayans.

It also becomes evident as to why the Mayans put such an emphasis on this specific date some 3,000 years into their future...and as to why they perceived such a place and time to be not only unique but also **SACRED!**

# VIII

## Footprints
## in the
## Sand

# Our Planet's Ocean Floors

# Footprints in the Sand

Recently, scientists have been able to look even deeper into our planet's ancient past. With new technologies they are now able to drill deep ice-core samples from the Antarctic regions and deep lava core samples from the depths of the Mid-Atlantic's Volcanic Ridge. They also obtain information from studying Earth's plate tectonics and the strata structure of various geological formations around the world.

From such studies paleontologists have put together NEW and STARTLING records of Earth's past climate and changes dating back to over 542 million years! What they have discovered is factual evidence for MASS EXTINCTIONS that have taken place on our planet over the millenniums. But, the startling evidence is that these ancient paleontological records show that such extinctions occur in a very precise, regular cyclical and reoccurring patterns...and proving that MINOR EXTINCTIONS occur about every 65 million years, with MAJOR EXTINCTIONS about every 250 million years. They have been able to trace this cyclical, repetitive pattern back to over 542 million years.

Since we estimate that our Galactic Year, one complete solar orbit around the Milky Way Galaxy, is somewhere between 225 to 250 million years. Scientists theorize that these cyclical extinctions are in direct correlation to each quarter-point in our cycle. At each of these points our solar system is believed to bob up and down vertically across our dense Galactic Disc.

This up-and-down motion periodically exposes the Earth (and all planets within our Solar System) to a higher dosage of dangerous cosmic rays. Research has also suggested that this motion may effect our climate as the solar system passes through the giant hydrogen clouds concentrated in the galaxy's spiral arms. Some researches have said that these cloud formations could be dense enough to literally sprinkle the Earth's atmosphere with dust, blocking out sunlight and leading to an immediate cooling of the planet. In addition, it is suggested that the gravitational pull of the clouds may dislodge comets from their spherical halo surrounding the solar system and send them literally crashing into Earth and her neighboring planets. An impact from a comet or meteor would further upset biodiversity and could, in and of itself, lead to extinctions. And, as if that were not enough, yet another possible problem may be the fact that the density of the hydrogen clouds could compress the solar wind, which shields our solar system from the deadly cosmic rays floating through our galaxy. (Note: In the following chapters you will find evidence that our current solar wind has quite recently been greatly compressed.) These cosmic rays are a regular part of not only our galactic environment but the entire universe as we know it. They are comprised of charged particles that have been accelerated to extremely high energies by supernova explosions.

Normally, our environment is such that these rays are continuously deflected away from our planet...but with the solar wind being compressed our solar shield is effectively down ~ thereby allowing these rays to leak into our planet's atmosphere, spurring the formation of clouds, cooling the planet, and destroying the ozone layer. Species would be killed off not only by absorbing these deadly ultraviolet rays, but also by other environmental changes which often lead to global cooling and ice ages.

**Image Credit: NASA ~ Chandra X  Gamma Ray Burst GRB 050709** "Right now a dying star in the constellation Sagittarius, WR104, may be on the verge of "going supernova". It's 8,000 light-years away, fortunately not close enough to vaporize us, but still within range to take out half the ozone layer and generate an Ordovician-like extinction. Oliver Reiser, a friend of Einstein's, wrote in the 1970's that evolution on earth was likely spurred by these occasional death zaps of interstellar radiation. Moreover, in his book The Intent of Creation, he theorized that cosmic rays routinely hit the Earth at intervals determined by the 26,000-year precession cycle. Unlike an asteroid or comet a GRB travels at the speed of light, precluding any warning of it's coming.

In addition, scientists have found that most of Earth's WORST extinctions have occurred when our solar system was at it's most northerly point in it's cycle on it's approach towards the Galactic Center. During this time our Sun and her planets stretch up to about 230 light years above the Galactic Plane, a point at which scientists theorize even more cosmic rays are entering into the Earth's atmosphere, killing off even more species. Our December 21st date in 2012 puts us precisely within this northerly location...as it is also the date of our Winter Equinox.

In the same way that the protective shield from the Solar Wind deflects and shields the Earth and her neighbors from incoming cosmic rays, the other stars within our galaxy also serve to produce a wind of charged particles whose magnetic fields deflect incoming cosmic rays from beyond our galaxy.

**...But....**

> **...as if that were not enough...**

> **... take this into consideration....**

....The entire Milky Way Galaxy is moving due north at 200 kilometers per second towards a galaxy called the Virgo Cluster. This movement compresses the galactic wind, allowing higher levels of potentially life-harming extra-galactic cosmic rays to reach directly into our solar system and eventually hit the Earth's surface. Scientists say that these periods of cosmic ray influxes match lows in biodiversity so well that the two factors have a one in a ONE-in-a-MILLION chance of BEING a COINCIDENCE.

# Earth Files:

A brief overview of Earth's paleontological history over the past 250 million years...

## PERMIAN EXTINCTION

## ~ 250 MILLION YEARS Ago

We begin with The Permian Age which ended approximately 250 million years ago. It was and is still considered to be the most MAJOR, WORST EXTINCTION known to have taken place on our planet. In fact, the effects were so devastating as to literally BREAK the surface of our planet

# ONE Continent~ONE OCEAN

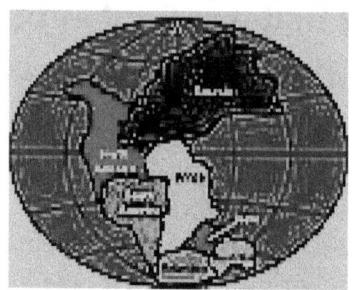

## The Great Pangaea
## 250,000,000 Million Years Ago

into pieces ~ which we know study as Plate Tectonics. Before the Permian Extinction our planet was made up of one major Super-Continent called the Pangaea, and one continuous ocean.

## Pangaea to Present

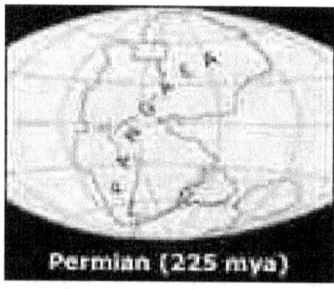

Permian (225 mya)

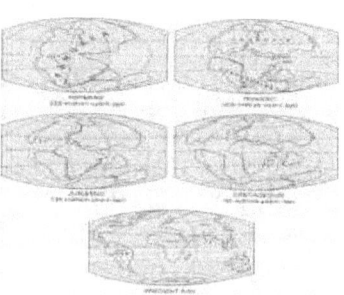

The event that took place must have been a geographical pole shift because there is no other known force strong enough to break apart the entire surface of a planet.

Immediately after that our surface was broken into the many continents and sub-continents that we see today. Since then, the continents have been drifting apart from one another, very slowly, over the past 250 million years. This is a phenomenon commonly referred to as continental drift. It is also known that during the Permian Extinction there was an enormous increase in volcanism within the Siberian Traps, leading to a loss of oxygen within the seas. **Great PERMIAN EXTINCTION~The Death Toll**: 95% of all species - broken down into 84% of all marine life & 70% of all land species including plants, mammals and insects.

# TRIASSIC EXTINCTION
## ~ 200 MILLION YEARS Ago

There is evidence that leads scientists to theorize that this extinction was most likely caused by massive floods of lava erupting from the central or mid-Atlantic Volcanic Ridge...and was also the event that triggered the opening of the Atlantic Ocean as we know it today. The volcanism may also have led to deadly global warming. Rocks from the eruptions are now found in the eastern United States, eastern Brazil, Southwestern United States (Arizona), Northern Africa and Spain. **The Death Toll:** 22% of all marine life, & 52% of marine genera (all sea life including vegetation, etc.) Vertebrate death remain unknown.


## JURASSIC EXTINCTION

## ~130 Million Years Ago

The Jurassic extinction occurred somewhere between 60 and 65 million years after the Permian Extinction, the greatest extinction event in the era of animal life on our planet. The cause of this extinction remains unknown. Scientist theorize that it could have been a meteor impact that caused immediate eruptions from our volcanic traps, or massively sustained eruptions over the course of a million years. Volcanic eruptions could also have triggered secondary effects such as global warming or cooling, which could also have led to a massive release of methane from the ocean floors. Methane is a highly poisonous gas that would immediately destroy all life forms on the planet. The main problem in determining the most likely cause behind this extinction is that the ocean crust recycles itself about every 50 million years...completely erasing evidence left behind by any major meteor or comet impacts. It is known, however, that the northern portion of the current-day Mid-Atlantic Ridge as well as the volcanic formations of the Central Atlantic were formed around this time.

There is another rarity located in a geologically unusual area of the southwestern United States. It is known as The Petrified Forest. This very unique area was also formed around this time. **Scientists believe that there are only a couple of events massive enough to have led to the instant petrification of this forest....one being an enormous ocean wave blasting over the area,** and the second being the aftermath of a massive meteor impact. **The Death Toll:** 30% of all marine sea life, 20% of all marine plant life & 50% of all land species.

# CRETACEOUS EXTINCTION
## ~ 250 MILLION YEARS Ago

There are several events that led to this most recent extinction which took place around 65 million years ago. Some argue that the extinction was caused by flood-like volcanic eruptions of basalt lava from the Indias Deccan Traps...leading to gradual climate change on the planet.

There is also evidence that the extinction was either caused or aggravated by a several-mile-wide asteroid that impacted the planet near the Gulf of Mexico, creating the Chicxilub crater now hidden on the Yucatan Peninsula and beneath the Gulf of Mexico. **The Death Toll:** 47% of marine genera , 16% of marine species & 18% of land vertebrate families including the dinosaurs.

# EARTH   FILE   SUMMARY

**Permian Period**    >    250 Million Years Ago
**= Minus 250 M**


**Triassic Period**    >    195-200 Million Years Ago
**= Minus 60 - 65 M**


**Jurassic Period**    >    130 Million Years Ago
**= Minus approx. 65 M**


**Cretaceous  Period**   >   65 Million Years Ago

**= Minus 65 M**

**Present Day**         >     65 Million Years to Date
**= Minus 65**


As of the date of this writing ~ **it has been**

## 250 – 250 MILLION YEARS

since the Great Permian Extinction,

&

## 65 million years since our last
## Secondary Extinction!

The above numbers look eerily familiar.  Roughly 250 million years ago, at the end of our previous Galactic Year and crossing the Galactic Center, we see the WORST extinction EVER to impact our planet.  The event was so powerful as to literally break apart the surface structure of our planet from one giant super-continent to the beginnings of the continents we see today. The only phenomenon that scientists agree could create such an enormous JOLT is that of an immediate GEOGRAPHICAL POLE SHIFT!

From there, we see the cyclical nature of mass extinctions occurring at approximately 65 million year intervals.  These intervals are also in line with the four major outpoints of our galactic orbit...or in Earth years what we refer to as the Spring, Summer, Fall and Winter Equinoxes.

...And, lastly...the fact that it has been around 250 million years to date that we last crossed our galactic center...and 65 million years since our last MINOR extinction...a very sobering thought as we approach our upcoming, fateful destiny with 2012

# Ancient Mayan Artifact
# The Mayan Calendar

# IX

# Back to the Future ...And Into The Past

# STRANGE FACTS

## In March 2009:

~Arizona becomes the 21st state to
DECLARE SOVEREIGNTY over the NATION.


~The United States announces that global
warming has quickly exceeded expect-
ations & that massive melting of our ice
caps will create an oceanic rise of over
three feet, creating coastal devastations
across the planet.

# Back to the Future...
# And Into The Past

N ew technologies have, very recently, allowed scientists to delve deeper than ever before into key areas of the Earth's surface....revealing startling, new information about our planet's ancient past. The two most highly critical and amazing pieces of evidence found were:

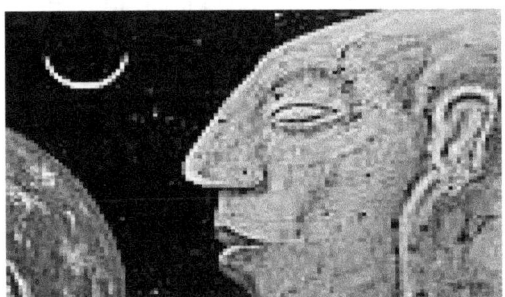

Layers of COSMIC DUST found in deep Antarctic ice core samples. Not only do these "layers" match up perfectly with our planetary Extinction Records, but the dust was also found to occur REPEATEDLY ~ over and over again ~ and in very PRECISE, CYCLICAL PATTERNS.

A very recent and STARTLING find has been made available to us as a result of brand new technologies enabling scientists to drill deeply into the Antarctic ice and obtain extremely deep ice core samples for the first time.

The first piece of startling evidence they unearthed was the discovery of ancient MAMMOTHS found buried deep within the ice core. Not only was this a shocking surprise as it proves that the Antarctic was once tropical and teeming with life, but the giant mammoths were literally FLASH FROZEN in time. They were unearthed perfectly preserved, and, frozen so quickly that their last bite of juicy foliage was found intact and PRESERVED within their jaws! These unfortunate creatures had been flash frozen so quickly they were unable to swallow their final morsel of food!

## Back to the future...and into the past...

...We were able to uncover information that is CRITICAL as we approach 2012. We all remember the paleontological Earth records that we were taught in grade school science class, but we probably never noticed that these records of past extinctions on our planet were both CYCLICAL, and also REOCCURRING! Geologists and paleontologists have been able to trace Earth's history back almost 750 million years, finding evidence that further confirms these precise patterns back even further into the ancient history, the past millenniums of our planet.

## Ancient Mayan

## Stone Artifact

As the above data shows, our planet dances across the millenniums with a "brief stop" about every 65 million years and a "grand stop" about every 250 million years. Is it merely a coincidence that these abrupt changes match precisely with our Sun's 250 million year orbit throughout the Milky Way ~ with the most MAJOR extinctions on our planet taking place on a regular, cyclical basis every 250 million years? In addition, archeological data also shows that our extinction events are  most severe when our planet is at it's most northerly point within it's orbit at the point of crossing the Galactic Equator.  This would most likely be during our winter months when we are at our furthest point from the Sun, and at our Winter Equinox.

December 21st of 2012 places us precisely in this position ~ as this is the exact time during which our Sun and her planets cross through the Galactic Center plane. So it is critical that we consider all data with the intent of determining whether or not we are ~ in fact ~ ending one 250 million year Solar Orbit and beginning a brand new Solar Year.  It is also important to note that scientists have traced this 250 million year extinction cycle back two cycles ~ to about 750 million years.

Then, almost exactly every 65 million years, Earth suffers another set of extinctions.  Perhaps we can equate these to our spring, summer, fall and winter equinoxes....as these would be the four equinoxes of the Sun's 250 million year orbit.  These secondary extinctions are also cyclical and reoccurring, but much less severe in nature, as can be seen by the above table.

# Cyclical & Reoccurring GAPS written into Earth's History

When all of the pieces of this giant puzzle are put together, what we have is a recorded pattern  of GAPS...CYCLICAL GAPS...woven intricately and precisely into Earth's past!  These GAPS reveal stunningly sharp and extreme changes on our planet's surface, changes that have taken place very quickly, within merely seconds or perhaps minutes! Such abrupt changes tend to virtually wipe-out or EXTINCT most life forms in existence at that period in time....and these repetitive, cyclical GAPS continue back into time as far as we are able to observe with present-day technologies.

# COSMIC DUST

The existence of Cosmic Dust layered into deep ice core samples from the Antarctica have only recently been discovered.  Deep ice core samples present the most PURE and PRECISE scientific evidence of our planet and it's atmosphere at the point in time during which they were frozen.  These samples are 100% preserved at the moment of their freezing and remain literally PURE and untainted by time.

The layers of COSMIC DUST found within these
deep ice core samples show that whatever caused
the dust to fall to the ground happened very quickly
and lasted for a very short period of time ~ perhaps
only moments, days or weeks.  But, whatever the
cause, what had taken place at the time the dust fell
was devastating to all life on our planet, and weirdly
enough, these samples  match precisely with all of
our other paleontological records of major Earth
extinctions.

The mystery left for scientists decipher is clear.  Is
this,  in fact, COSMIC DUST?  And, if so, how can it
be proven?  Secondly, was it the dust itself that led to
the extinctions...or was the dust merely evidence left
behind like traces of DNA evidence remaining at a
crime scene?

Both the discovery and the analysis of cosmic dust
layers within deep ice core samples are accredited to
one scientist by the name of  Dr. Paul LaViolette.  Dr.
LaViolette holds 9 degrees in physics from Johns
Hopkins, an MBA from the University of Chicago, and
a PHD from Portland State University.  He was the
first to discover high concentrations of Cosmic Dust
within the Ice Age Polar Sheet.  Based upon his
findings, he made predictions about the entry of
interstellar dust into our solar system 10 years before
it's confirmation in 1993 by data from the IRAS and
Ulysses Spacecrafts and by radar observatories in
New Zealand.

And, since the discovery of the fact that Cosmic Dust is now streaming into our heliosphere, the amount of dust has been steadily increasing....by about three times since 2001.

Dr. LaViolette also conducted ice core studies from different depths in order to be more precise...and he found that the cosmic dust layers within the samples matched the times of most known animal extinctions on Earth, including the time of the dinosaur's mass extinction!

The next puzzling aspect to the presence of the cosmic dust is HOW it made it's way into Earth's atmosphere? The Earth has always been protected by the Sun's heliosphere, which not only deflects cosmic rays, but also other things whizzing through our galaxy...things such as comets, meteors, asteroids, space rocks, etc. The answer becomes quickly apparent. The DUST did not come to Earth...but the Earth, along with the Sun and the rest of her planets, went to it!

As we discussed previously, the Galactic Center region is an area that is densely littered with dust. In fact, there is so much dust that this region has always appeared to be dark and cloudy and is the reason that the ancient Mayans referred to it as the "Dark Rift." So...clearly we are bombarded with Cosmic Dust each time we pass through the Galactic Center, and each time we also experience one or more major events which result in planetary upheavals and extinctions.

In fact, it was at one of these crossings that our planet took on it's current axis with the North and South poles presently sitting in a "tilted" position...and off of our axis with the Sun...thereby creating the four seasons and diversity of climates we have today. And, it was at one of these crossings that the mammoths were flash-frozen in time in the Antractica. Scientists have also been able to prove that both our North and South Polar Regions were once tropical paradises, teeming with life...and home to many exotic species of tropical plant and animal life forms. But...in order for this to have taken place...geologists and astrophysicists state that our planet must have sat at a PERFECT EQUILIBRIUM to the Sun. Such an equilibrium would make our entire planet one season all year long and is what the scientific community refers to as "Perpetuum Vernnum" or "Eternal Spring."

Then something quite startling happened! Very suddenly, not over a period of years or even days, but over a period of minutes or seconds...EVERYTHING CHANGED! And...it changed so quickly that many forms of life were literally flash-frozen in time! There is only ONE EVENT that scientists know of that would be able to cause such EXTREME change so abruptly ~ and that event is called a **GEOGRAPHICAL POLE SHIFT!**

During a geographical pole shift our planet "takes a 180" and literally FLIPS OVER in space with the center core of our planet counter-rotating against the surface layer. Much like a top spinning on a table our planet very quickly reorients itself in space. To Earth this is a normal event...no big deal...life goes on. But to those inhabitants on the surface of the planet life is abruptly and dramatically changed. Suddenly the Sun itself is in a different location. It may rise in the North or the West. The topography of the planet is also drastically different. There may be oceans covering entire continents, and new continents may appear where there was once only ocean. Likewise, the Earth's climate is immediately and dramatically altered. The worse part however, is that these IMMEDIATE and CATASTROPHIC changes are so INTENSE that they quickly lead to mass extinctions.

And the dust? The layers of dust found in the ice core samples are merely left-over remnants from the pole shift...that is when the pole shift occurs during a crossing of our Galactic Center. As we remember the center region of our galaxy is very thickly and densely filled with Cosmic Dust...a dust so thick that we are unable to observe anything within the center region, even with our best space telescopes.

This also explains the giant mammoths that were literally flash frozen in time in the Antarctica...and the fact that the arctic was tropical and teeming with life one second and then a frozen tundra the next. A geographical pole shift happens SUDDENLY, and is over within merely a few moments. But, during that brief time everything has changed!

# So....IT DID HAPPEN....

# AND

# It happens REGULARLY in PRECISE, CYCLICAL PATTERNS!!

Many astrophysicists believe that we will experience one or more geographical pole shifts during our upcoming galactic crossing in 2012. Most definitely, not the best of news...but KNOWLEDGE brings us one step further to survival! And...finally...we have the proof to solve perhaps the most important mystery of life on this wonderfully beautiful planet we call home!

If you would like to find out what it would be like to live through a "Day in the Life of a Pole Shift"

&

Experience an Amazing Journey through time...

Please look for the 2nd the 3rd books in "Trilogy 2012"

...available soon!

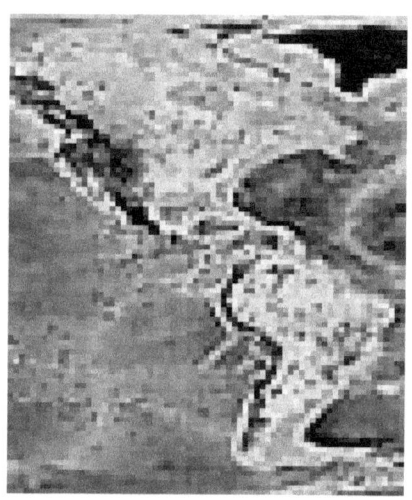

Map of Earthquake Areas in North & South America

# X

# The Ground Beneath Your Feet

# Ancient MAYAN Depiction

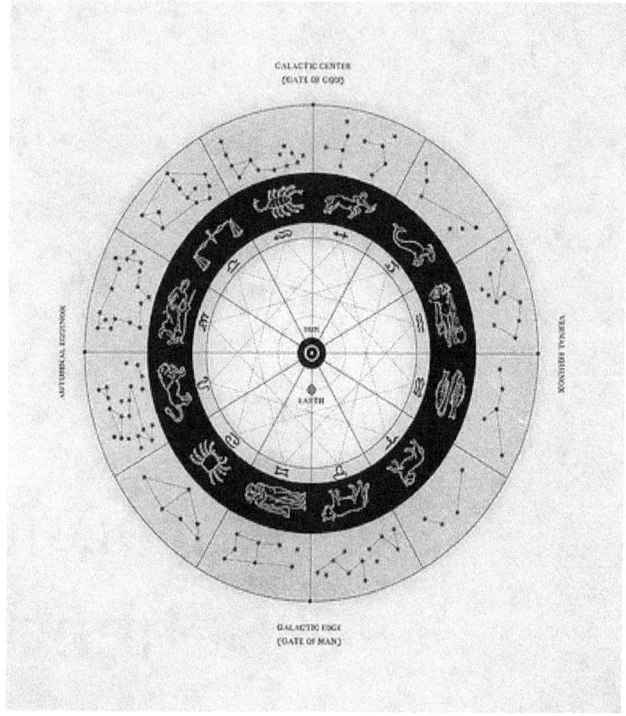

## of Galactic Center and our Solar Orbit

# The Ground Beneath Your Feet

The Earth itself is considered to be a living organism." As such, it is constantly active, changing, reforming and renewing itself. Perhaps the force that imposes the greatest threat to mankind is the endless movement of our planet's tectonic plates. Such movements very often result in volcanic eruptions, earthquakes, tsunami's...and, in very rare instances, the total annihilation of entire continents! As we will discover within the next few pages...plate movement has been becoming increasingly active over the past 2 or so years. As a result we have experienced a huge increase in volcanic activity, earthquakes, and tsunamis worldwide. It is wise to know the plate configuration of our planet...and, most importantly, the plate boundaries in your own particular area. These would be the areas most heavily impacted by any future Earth changes.

At some point in the very distant past ~ at the time of the Permian Extinction event about 250 million years ago ~ the surface layer of our planet split into 16 separate pieces or plates. All of these pieces fit together perfectly, much like an enormous jigsaw puzzle. These plates are very important to life on Earth because they are the MAIN FORCE that shapes our  planet's surface, and they are continually in movement with and against each other.

Our planet's present-day continental and oceanic plates include: the Eurasian plate, Australian plate, Philippine plate, Pacific plate, Juan De Fuca plate, Nazca plate, Cocos plate, North American plate, Caribbean plate, South American plate, African plate, Arabian plate, Indian plate,  Antarctic plate, and the Scotia plate.  Many of these major plates also contain smaller sub-plates.

This outer layer of the Earth's surface that consists of this tectonic plate structure is called the lithosphere.  The plates are made of rock, but the rock is, in general, light-weight compared with the denser, fluid underneath.  This allows the plates to literally "float" on top of the denser material.

Movements deep within the Earth, which carry heat from the hot interior to the cooler surface, cause the plates to move very slowly on the surface, about 2 inches per year. Very interesting, and sometimes strange, things happen at the edges of plates.  Let's put it this way...the intersection of two plates would NOT be the safest place on the planet to live!

# Continental Plate Boundaries ...

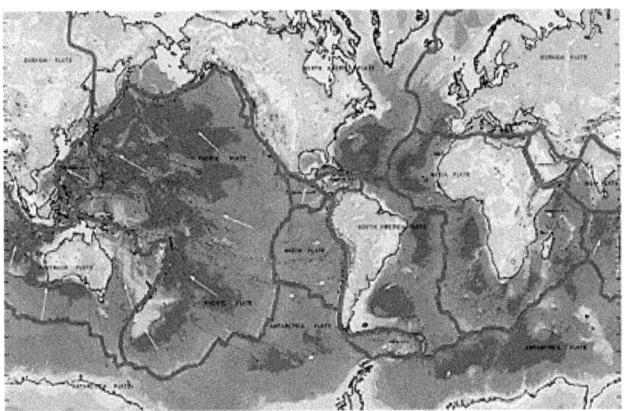

Three types of movements happen when two or more plates come together. First, forces from deep within the Earth cause them to literally crash together, forming what are called **Subduction Zones**. At a subduction zone one plate is forced back down underneath the second plate, where it is then pushed deep into the Earth and eventually melted. The oceanic plates (being less dense) normally sub-duct underneath the much denser continental plates. When two oceanic plates collide, one may be pushed under the other causing magma from within the Earth's mantle to rises up. This action leads to the formation volcanoes along this vicinity. Melted crust then rises back up to the surface where it forms volcanoes and islands. Thus the formation of some volcanoes, mountains, and islands is connected to the process of subduction and continental drift.  Most subduction zones are located along the edges of the continents, or the continental shelves where the land meets the water.

The second type of action occurs when plates "pull away" from each other creating **Spreading Ridges.** Most of today's spreading ridges are found in the central portion of our sea floor oceans.  Seafloor spreading is the movement of two oceanic plates being forced away from each other.  This is referred to as a divergent plate boundary and results in the formation of new oceanic crust being created from magma that comes from deep within the Earth's mantle.  This action occurs along what are called mid-oceanic ridges.  When two continental plates collide the Earth's crust is compressed and pushed upwards forming mountain ranges.

And, last but not least, there are the large and familiar **Fault Lines** that form when plates slide past each other. When two plates move sideways against each other they form a transform plate boundary. Needless to say this action creates a tremendous amount of friction and  makes the movement extremely jerky. The plates slip, then stick, as the friction and pressure builds up to incredible levels. When the pressure is finally released, the plates suddenly jerk apart creating very strong and often deadly earthquakes.

# Map of Plate Boundaries

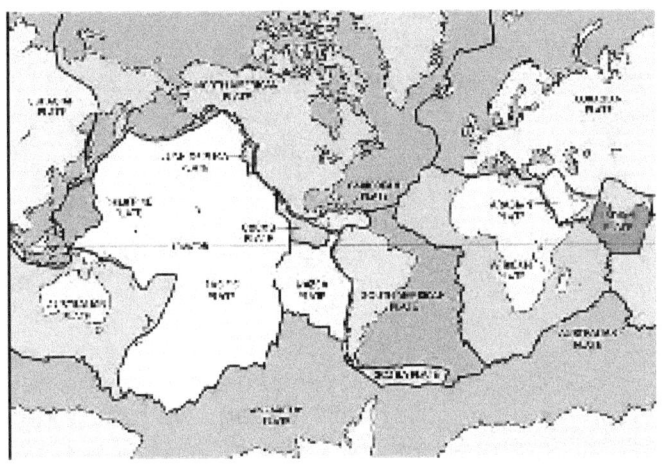

# CYCLES of the GREAT MAYAN CALENDAR

| Cycle | Composed of | Total Days | Years (approx.) |
|---|---|---|---|
| kin | | 1 | |
| uinal | 20 kin | 20 | |
| tun | 18 uinal | 360 | 0.986 |
| katun | 20 tun | 7200 | 19.7 |
| baktun | 20 katun | 144,000 | 394.3 |
| pictun | 20 baktun | 2,880,000 | 7,885 |
| calabtun | 20 piktun | 57,600,000 | 157,704 |
| kinchiltun | 20 calabtun | 1,152,000,000 | 3,154,071 |
| alautun | 20 kinchiltun | 23,040,000,000 | 63,081,429 |

The Ancient Mayans developed a calendar system far more advanced than any we have today. With this system they tracked the cycles of our planet and solar system as we traveled our course throughout the Milky Way Galaxy ~ tracking our orbital location up to over 63 million years.

All of our planet's top scientists and astrophysicists ~ including NASA ~ hail their system for it's detail and on-target accuracy that continues almost 5,000 years into the future. The fact that the Alautun cycle is composted of 63,081,429 years is NOT a random coincidence. Scientists estimate our Galactic Orbit to be somewhere around 250 MILLION YEARS ~ making each Alautun a 1/4 point ~ or equinox ~ within our orbit. These points are highly significant since geological records show that it is at each of these equinoxes that we experience our most severe extinctions. The ancient Mayans were most obviously aware of this fact as well!

# XI

# Earth in ABRUPT Hyper-Dimensional Change

Earth changes began in the 1970's and have been growing exponentially ever since. The scientific community  recently announced that **our planet is now well into her 6th Extinction** ...but they can only speculate as to why.

## WHY is This HAPPENING?

It is most certainly EVIDENT to all of us currently residing on planet Earth that we are in the midst of DRAMATIC CLIMATE CHANGE...and that something

VERY, VERY UNUSUAL and enormous in scope is taking place on our planet.  Paleontological evidence from our planet's ancient past shows that the Earth undergoes **massive changes** in a **regular, cyclical pattern** that is **alarmingly consistent...almost as if it were a part of the "grande plan" for the Earth to experience a purification, cleansing, renewal, regeneration and REJUVENATION that is predesigned to take place every so often on a regularly occurring basis.**

## Earth's Pattern of CYCLICAL CHANGES:

EVERY 250 MILLION YEARS Earth undergoes MASSIVE Changes & MASSIVE Extinctions.

EVERY 62 – 65 MILLION YEARS Earth undergoes SEVERE CHANGES & EXTINCTIONS of a Lesser Magnitude.

## CRITICAL NOTE:

As of the date of this writing it has been ~ 250 Million Years since the Great Permian Extinction.

# AND

## ~ 65 Million Years since our last Secondary Extinction!

## SO.....

Is what we are experiencing now merely the "scheduled" REJUVENATION of our beloved planet? Is Mother Earth merely "checking-in" for her routine "SPA-REJUVENATION?" We all know women, and...much like our planet...not one of us likes to look our age!

And...is it merely by coincidence that our 250 Million Year Extinctions correspond exactly with our Sun's Galactic Year, bringing us back right smack on target with our Galactic Center? Is it also by coincidence that our Secondary Extinctions correspond to each of the 4 "equinoxes" of our Galactic Orbit...much like the Spring, Summer and Fall Equinox's of our orbit around the Sun? Scientists have also recently discovered that the changes taking place on Earth are happening on every planet within our Solar System...including the Sun herself! In fact, our planet has not experienced even a fraction of the changes seen on our neighboring planets. It is almost as if we are being protected by some unseen force! And, if that is the case...maybe, just maybe...we will be able to survive our forth coming destiny with 2012!

With that positive thought in mind, what types of changes are we currently experiencing? World-renowned physicist, Dr. Brooks Agnew, believes that the signs of massive Earth changes are already here, happening right now, and are all around us. Dr. Agnew also believes that what we see happening now are the "undertow affects" of approaching our Galactic Center, and that these "undertow affects" will become stronger and stronger as we approach that crossing.

Many physicists agree with Dr. Agnew that these affects will become stronger and stronger until we are finally hit with the **big one** as we cross our Galactic Equator. Others believe that the FORCES at PLAY will come upon us in one huge BLAST at the time of our Galactic Crossing in 2012. And, there are those few within the scientific community whom believe that little or nothing will happen.

# ...BUT...

## There is

# ONE SIMPLE FACT

# Everyone
# AGREES UPON!

The fact that we are currently
## ON APPROACH
to and
# WILL CROSS
our
# GALACTIC CENTER
At
# PRECISELY
## 11:11:00:00 AM
## on
## 12:21:12:11:11:00:00
### December 21st, 2012!

## Earth's Winter Solstice

## GALACTIC ALIGNMENT
## 21st December 2012 AD

# Earth in Abrupt
# **HYPER-DIMENSIONAL**
# <u>Climate Change:</u>

# II.0

# Earth in Abrupt Hyper-Dimensional Climate Change

In December of 2008 the U.S. Climate Change Science Program submitted a report on "Abrupt Climate Change" to the President and the Congress. The report was prepared by the U.S. Geological Survey, the National Oceanic and Atmospheric Administration (NOAA), and the National Science Foundation (NSF).

# The Report Addressed
## 4 Areas of MAJOR CONCERN:

___

RAPID changes in our glaciers, ice sheets & sea levels.

___

Widespread changes to our hydrological cycle.

___

Abrupt changes in "overturning circulation" throughout the Atlantic.

(The northward flow of warm, salty water in the upper layers of the ocean.)

___

Rapid release of the methane trapped in permafrost and on continental margins.

The presentation went on to discuss planet-wide changes that have taken place over the past 30 & 50 years:

# Over the PAST 50 YEARS:

———

There has been a  large drying trend over much of the Northern Hemisphere, with the opposite trend in the Eastern, North and South Americas.

# Over the PAST 30 Years:

———

There has been a 75% Increase in the number of STRONG Category 4 HURRICANES across the globe - with the overall potential DESTRUCTIVENESS on an UPWARD TREND.  The number of hurricanes in the North Atlantic has also been above normal over the past 11 years, and reached  a record-breaking season in 2005.

———

Extremely DRY LAND across the globe  has more than DOUBLED.

# Extremely UNUSUAL Trends:

## - ALARMINGLY HIGH LEVELS of CO2 in our Atmosphere

CO2 has now reached a record high for the past 2 MILLION YEARS! Even more significantly, it has done so at an exceptionally fast rate. The report stated that, "The concentration of CO2 is now known accurately for the past 650,000 years from Antarctic Ice Core samples. During this time, CO2 concentration varied from 180 ppm to 300 ppm. But over the past century, CO2 has rapidly increased to an unprecedented 379 ppm. A rise such as this would normally take MANY MILLIONS OF YEARS to occur!

## And, the GLOBAL WARMING Trend is Rapidly Increasing.

## The report ended with an unsettling conclusion:

"Further back in time, beyond ice core data, evidence from sediment cores and other archives do not resolve changes as rapid as the present-day warming trend. Hence, although large climate changes have occurred in the past, there is no evidence that this rate of change was matched by any comparable global temperature increase **over the last 50 million years!**"

# 11.1

## All Life on Planet Earth

The Interrnational Union for the Conservation of Nature (IUCN) has named 2010 as the International Year of Biodiversity. On February 18th of 2010 the IUCN made the startling announcement that "one-half of the world's primates are currently on the threshold of extinction!"

The IUCN was founded in 1948. Today, it is the world's oldest and largest global environmental network, working in more than countries and with over 1,000 government and non-government agencies worldwide.

The 2008 IUCN Red List Update includes species which are currently considered Globally Threatened or Extinct. And...if you don't have tears in your eyes already...the even more startling, tear-jerking, aspect of all of this is that the variety of species existing today are a product of **3 & 1/2 BILLION YEARS of EVOLUTION!**

**Following is a brief summary of the present Extinction Rate (by %) & by species of the date of this writing:**

> 22% of all known mammal species

> 31% of all amphibians

> 14% of all birds

> 28% of all known reptiles

> 37% of all freshwater fish

> 70% of all plants

> 35% of all invertebrates

> 27% of warm-water, reef-building corals

> . 12% or all 161 grouper species

> 16% of all 1,280 species of freshwater crabs.

**New species added to the list in 2008 included:**

−        14 Tarantulas from India

−        3 Orchids from the Americas

−        the Bumblebee which has mysteriously taken a dramatic decline in North America within the past couple of years.

# 11.2

# Loss of Earth's Biodiversity

The loss of biodiversity, alongside with climate change, is one of the most CRITICAL CHALLENGES of our time!  Our planet's ecosystem has always had a very delicate and intricate design, much like that of a giant "free-standing" puzzle structure with millions and millions of tiny, intricately-fitting pieces.  Each piece links together with the next...forming what Native Americans refer to as The Circle of Life.  BUT...the system is so intricately DELICATE, that when even the tiniest of pieces are removed or destroyed, the ENTIRE structure begins to collapse in upon itself.

Every plant and animal...down to even the smallest of microbial life forms not even visible to the naked eye...are intricately and forever connected to their brothers. .All of God's life forms "holding hands," intricately and minutely taking shape to form this PRECISELY PERFECT and SACRED "Circle of Life." Scientists refer to it as being a part of the overall GRANDE DESIGN of our universe...a DESIGN so PERFECT that it has forced even the most cynical of scientists to admit the fact that there **IS** an **Intelligent Design** behind everything...all creation . ..within our Universe!

So...scientists FINALLY admit that there IS an Intelligent Design to our world...proving that there must have been a Supreme Force, a CREATOR! No matter what we call this intelligence: The Creator, God, The God Force, Our Higher Power...or by whatever name we wish to use...a CREATOR does EXIST! And, by DESIGN, that creator made the biodiversity of our planet  as essential to LIFE as the sunlight, the water, and the air that we breath. Nothing can exist without it...AND...it must remain intact and intricately PERFECT in order to exist at all...and...in order for ANY ONE of US to EXIST at all!

# Can Loss of Biodiversity cause Mass Extinctions?

Scientists, biologists and organizations worldwide agree upon the fact that our planets loss of biodiversity puts us in extreme danger of mass extinction. In fact, not since the dinosaurs died out has Earth suffered such a rapid loss of plants, animals, and ecosystems. Our destruction of nature could be even more costly than our greenhouse gas emissions.

The Circle of Life...is a minutely-perfect & SACRED...biological, biochemical and electromagnetic interaction amongst all of God's Creations...

The Millennium Ecosystem Assessment shows that human actions often lead to irreversible losses in terms of diversity of life on Earth and these losses have been more rapid in the past 50 years than ever before in human history.

Our very wise, and dearly loved, Native Americans have always understood and held extreme reverence for the Circle of Life, and were perhaps the very people who originally coined the phrase. But...to the rest of us...this has most probably been a thought that has never even entered into our busy, hectic daily lives.

# "The Circle of Life"

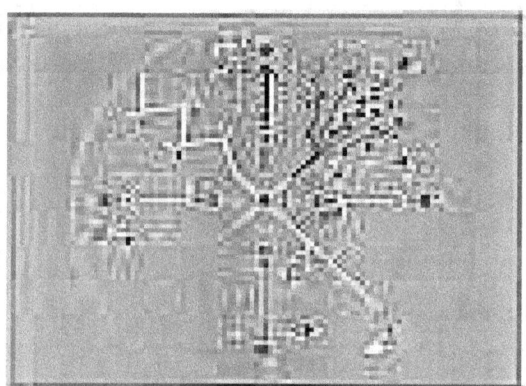

## The HOPI Nation

Biodiversity is, quite simply, the **"Circle of Life."** All life on our planet, our entire ecosystem, has a grand and extremely intricate design.  From the smallest microorganism in the depths of the oceans to the largest tree and animal on the planet....each life form is built upon the one beneath it.  It is much like a very delicate circle with billions of tiny interacting gears.  In it's designed state it can run forever with basically no effort.  But...should one tiniest of tiny links be taken away...the entire system will very quickly begin breaking apart.  Each piece of our circle of life is mandatory and critical. When one link is removed a domino affect begins immediately and continues growing faster and faster like a snowball rolling down a freshly snowed mountain.

Biodiversity is everywhere, both on land and in water. It includes all organisms, from microscopic bacteria to more complex plants and animals. The United Nations currently estimates that we lose 3 species an hour to urbanization, deforestation, overfishing, climate change, and invasive species.

# HOPI Prophecy

# The Hopi Nation

Dedication: In Loving Memory
of my Beautiful White Wolf NATASHA

# SAVE A HONEYBEE...
# A Dear & Beautiful Creature
# STRUGGLING for LIFE!

If you just happened to be on a gorgeous, Caribbean beach in Negril Beach, Jamaica a couple of winters ago...you may have noticed one of the many "beach bunnies" making regular, frantic trips from the ocean up to the beach and into the deep foliage, where she appeared to be placing small creatures that she rescued from the sea. And then as she approached her lounge on the beach, you would have noticed that there were at least two small, coyote-looking Jamaican dogs curled up under her thatched roof umbrella...YES...I was subject to a number of strange looks...but AT LEAST I WASN'T TOPLESS! That was me, frantically saving these sweet, tiny drowning honeybees from the depths of the ocean. The strangest part of it was that it was only a few months before the Honeybee Disappearance Mystery began ~ but, I always make it a point to follow my instincts ~ no matter how strange they may seem to be. And the little Jamaican dogs...I seem to attract animals and wildlife of all types wherever I go. In fact, at my residence you never know what type of creatures may show up for unannounced visits...everything from flocks of ducks (and I do not live near any water) to an entire family of peacocks ~ all with their adorable faces pressed against my front window.

BACK to the HONEYBEE MYSTERY...Within the past few years scientists have been completely puzzled by the sudden collapse of almost all honeybee colonies worldwide. Many bee farmers and beekeepers have been forced out of business...but the mystery of what happened so suddenly, and why, is still at large.

Of course, we all love honey...but the loss of our beloved honeybees has much deeper and graver consequences for us all. For one thing, without the work of the honeybees to pollinate and cross pollinate many, many other species of plants, flowers and vegetables will cease to exist. In 2007 alone, the loss of pollination caused U.S. Agriculture to lose $15 BILLION DOLLARS! Industrial farmers have already been forced to reduce the diversity of many of their crops. This, in turn, reduces many other factors essential to crop development...thus leaving our food chain even more vulnerable. And...if our coral reefs do disappear within the near future...many, many species of fish and other sea life will also die off...and another $152 BILLION DOLLARS of annual revenues will go with them.

The following article from The Washington Post, April 21,1998 clearly sums up the seriousness of this problem:

## "Mass Extinction Underway, Majority of Biologists Say"

**Washington Post** Tuesday, April 21, 1998

**By Joby Warrick, Staff Writer**

"A majority of the nation's biologists are convinced that a "mass extinction" of plants and animals is underway that poses a major threat to humans in the next century, yet most Americans are only dimly aware of the problem, a poll says. The rapid disappearance of species was ranked as one of the planet's gravest environmental worries, surpassing pollution, global warming and the thinning of the ozone layer, according to the survey of 400 scientists commissioned by New York's American Museum of Natural History.

The poll's release yesterday comes on the heels of a groundbreaking study of plant diversity that concluded than at least one in eight known plant species is threatened with extinction. Although scientists are divided over the specific numbers, many believe that the rate of loss is greater now than at any time in history. "The speed at which species are being lost is much faster than any we've seen in the past ~

including those [extinctions] related to meteor
collisions," said Daniel Simberloff, a University of
Tennessee ecologist and prominent expert in
biological diversity who participated in the museum's
survey. [Note: the last mass extinction caused by a
meteor collision was that of the dinosaurs, 65 million
years ago.]

Most of his peers apparently agree. Nearly seven
out of 10 of the biologists polled said they believed a
"mass extinction" was underway, and an equal number
predicted that up to one-fifth of all living species
could disappear within 30 years. Nearly all attributed
the losses to human activity, especially the destruction
of plant and animal habitats. Among the dissenters,
some argue that there is not yet enough data to support
the view that a mass extinction is occurring. Many of
the estimates of species loss are extrapolations based
on the global destruction of rain forests and other rich
habitats.

Among non-scientists, meanwhile, the subject
appears to have made relatively little impression.
Sixty percent of the laymen polled professed little or
no familiarity with the concept of biological diversity,
and barely half ranked species loss as a "major threat."

The scientists interviewed in the Louis Harris poll were members of the Washington-based American Institute of Biological Sciences, a professional society of more than 5,000 scientists."

**UNCTAD/PRESS/IN/2010/001     12/01/10**

# 11.3

## The State of our Oceans

The ocean occupies 70% of the Earth's surface and is home to a large percentage of our planet's biodiversity. When volume is taken into consideration an even larger percentage of habitable space is affected.  There is an ever growing concern that a broad range of marine species are under risk of extinction and that the biodiversity of our oceans are experiencing potentially irreversible loss due to threats such as climate change, over fishing and coastal development.

We all know that life literally began in the oceans...and after millions of years slowly emerged onto land...a fact which makes it all too clear that the ecosystem of our oceans have **A DIRECT IMPACT on all life on Earth**. As with the planet as a whole, oceanic life is minutely sensitive to any type of change.  For example, the warming of the water by merely a couple of degrees is enough to throw all life in the ocean into a downward spiral.  Both global warming and the melting of our polar ice caps are presently creating such an affect.  To date, the most eminent area of concern is the loss of our precious and exotic corals...which are presently experiencing a fast and fated degradation.

−       **EARTH's TREASURED CORAL REEFS are DYING:**  Our planets most exotically beautiful and treasured coral reefs are dying all over the planet. Almost ½ of the worlds' reef building corals are now listed as Threatened or Near Threatened. Two major factors are influencing this major loss to our ocean's biodiversity:

# 1) Temperature induced bleaching on a wide geographic scale.

# 2) The increase in atmospheric CO2 to levels exceeding 320 ppm.

Both of the of the above factors have led to an increase in sea temperatures and disease, and an increase in the **frequency and duration of the bleaching of our coral reefs.**

—      **CO2 Levels also Affect Sea Life!** Scientists are extremely concerned that if CO2 levels rise to 459 ppm reefs will be in rapid and terminal decline world-wide...which will have an enormous effect on the already degrading state of our oceans biodiversity...and be followed by even more extinctions. In addition, there will be knock-off affects to many other ecosystems...and should CO2 levels reach 600 ppm...numerous domino effects will follow, affecting many other marine ecosystems. The scientists concluded in this report, "This is likely to have been the path of great mass extinctions of the past...and could put into full force the Earth's 6th Mass Extinction!

– 　　**SEA TEMPERATURES are RISING!** The sea temperature of the Antarctic Ocean rose by 3 percent in 2010 ~ a direct effect of the extreme melting of the Antarctica region including the ice caps, glaciers and glacial sheets. Scientists consider this seemingly low increase to be very EXTREME for the biodiversity and other functions of the oceanic currents and it has already lead to a SIGNIFICANT GLOBAL INCREASE IN OCEANIC TEMPERATURES WORLDWIDE!

– **MARINE MAMMALS THREATENED!** 25% of marine mammals are now threatened. Marine mammals include: whales, dolphins, porpoises, sea lions, walruses, sea otters, marine otters, manatees, dugongs and ~ LAST but not LEAST ~ our beloved POLAR BEARS! Major threats to marine mammals include water pollution, habitat loss from coastal development, loss of prey or other food sources, intensive hunting, and the combined effects of climate change.

– **SEABIRDS:** 27% of the world's seabirds are threatened. Major threats to seabirds include mortality fisheries, gill-nets and OIL SPILLS. The loss of breeding and habitat sites pose yet an additional threat.

– **Oceanic DEAD ZONES:** A very strange anomaly noticed by scientists over the past few years is what is

162

referred to as Oceanic DEAD ZONES. These areas occur throughout all of our oceans within various spots across the globe and are a complete mystery to scientists and biologists.

A dead zone is a patch of ocean that is literally DEAD, stagnant and 100% VOID of any forms of life, including microscopic and microbial life forms. A quick search on Google Earth will point out all Dead Zones currently known to exist. Biologists are actively studying these areas to determine their rates of growth and decline.

– **ISLANDS of TRASH:** Space telescopes that track the planet on an on-going, daily basis have also recently found large "islands" of trash and plastic which are believed to have recently emerged around the planet. As caretakers of this planet, this is one fact that we are most certainly accountable for...but at least one that can be easily fixed.

# Ancient Mayan

# Pyramid

# 11.4

## North Antarctic becomes an Ocean for the 1st Time in Human History

Earth changes went into ACCELERATION beginning in 2003. Since then we have experienced an unprecedented and EXTREMELY ACCELERATED melting of all of our ice caps, glaciers and ice sheets, Planet-Wide! And...to make matters even worse ...scientists have noted that our glaciers and ice caps are melting at a MUCH, MUCH FASTER rate than was expected!

**November 2003:** Scientists note that the melting of our ice caps and glacial sheets has created a snowball or domino affect...causing the remaining ice to melt much, much faster than was originally predicted...and DRAMATICALLY ACCELERATING the melting of all glaciers and ice sheets planet-wide.

**February 2004:** The Pentagon released a report on "Abrupt Climate Change" in order to address the loss of our ice caps along with other severe problems currently affecting our planet.

. **Summer 2004:** Then a few months later the shocking EVENT HAPPENED and a major unprecedented event occurred! It was the first time in the history of mankind on this planet that ALL of the ICE at the North Antarctic Pole melted. To the surprise of all of the greatest scientific minds in the world...our North Polar Ice Cap had become an ocean!

And...since then...both our North and South Polar Ice Caps, along with all of our continental glaciers and ice sheets, have been continuing to melt at ever growing rates.

**By 2010:** The sea temperature of the Antarctica had risen by a dangerous 3%.

As of the date of this writing, the massive melting of our planet's ice caps and glacial sheets has caused the sea temperature of the Antarctica to rise by 3%. This may appear to be a small increase, but it is HIGHLY SIGNIFICANT with respect to oceanic temperature, overturning, and circulation, and will adversely affect all oceanic sea life. And, the bad news is that it will also very quickly and significantly boost the speed of further melting and lead to even higher oceanic temperatures on a planetary basis!

# 11.5

## CO2

## Reaches

## 800,000 Year

## HIGH

One of the planet's leading oceanic scientists recently warned us that carbon dioxide threatens a mass extinction of all sea life! . Atmospheric $CO_2$ is now higher than the natural range of 180 to 300 ppm that has existed for the past 800,000 years...and...it is about 100 ppm higher than the 280 ppm that existed for 10,000 years before the start of the Industrial Revolution.

$CO_2$ continues to climb higher each year...at a pace that accelerates from decade to decade. Twenty years ago, in November 1988, atmospheric $CO_2$ was 349.99 ppm. Twenty years later, we are now able to mark that as the critical point in time, the time that we crossed "the line" between SAFE and DANGEROUS $CO_2$. In addition, researchers say that since the Industrial Revolution, $CO_2$ emissions have turned our oceans about 30% more acidic, and that they are now more acidic than they have been for the past 500,000 years. The problem is set to worsen as emissions of greenhouse gases increase throughout the 21st Century.

Dr. Carol Turley, Chairman for the recent Copenhagen Climate Change Congress, made some startling statements regarding the current state of our oceans:

"I am very worried for ocean ecosystems which are currently productive and diverse," Carol Turely told BBC News. "I believe we may be heading for a mass extinction, as this rate of change in the oceans hasn't been seen since the dinosaurs. It may have a major impact on food security. It really is imperative that we cut emissions of $CO_2$."

Dr. Turley, from Plymouth Marine Laboratory, went on to say that it is impossible to know how marine life will cope but she fears many species will not survive.

# CO2 Surges from 350ppm  to 384ppm  in Twenty Years

Mauna Loa Observatory

**Hawaii, USA** "Atmospheric CO2 was 388.63 parts per million (ppm) in the first month of 2010", according to scientific data released February 10, 2010, by the National Oceanic and Atmospheric Administration (NOAA) in the United States. Atmospheric CO2 was 386.92 ppm one year earlier in January 2009.

As can be seen by the following data, CO2 has increased by 73 ppm over the past 51 years! It began to increase in 1960, and went on to take a second dramatic increase over the course of that decade ~ ending with an overall increase of 9 ppm by 1970. Then for the next four decades it continued to increase ~ and is now at a staggering increase of 19.5 ppm for the decade ending in 2010.

| | |
|---|---|
| **1960 - 1970** | **9.0 ppm** |
| **1970 - 1980** | **13.0 ppm** |
| **1980 - 1990** | **16.0 ppm** |
| **1990 – 2000** | **17.0 ppm** |
| **2000 – 2010** | **19.5 ppm** |

The Mauna Loa Observatory began it's high-precision instrument readings in March of 1958.

169

Complete sets of monthly data for atmospheric $CO_2$ are available to the public at the Mauna Loa Observatory.

## CO2 Levels  from January 1959 – January 2010

| | |
|---|---|
| January 2010: | 388.63 parts per million (ppm) |
| January 2009: | 386.92 |
| January 2008: | 385.42 |
| January 2007: | 382.88 |
| January 2006: | 381.36 |
| January 2005: | 378.43 |
| January 2004: | 377.03 |
| January 2003: | 374.92 |
| January 2002: | 372.38 |
| January 2001: | 370.47 |
| January 2000: | 369.07 |
| January 1990: | 353.60 |
| January 1980 | 337.80 |
| January 1970 | 325.03 |
| January 1960 | 316.43 |
| January 1959 | 315.62 |

# 11.6

## Shake
## Rumble
## &
## Roll

INCREASE

in

EARTHQUAKES
by FREQUENCY &
INTENSITY

Scientists say that the apparent increase in earthquake activity may be linked to climate change, as the melting of glaciers and ice sheets directly induces tectonic activity. The reason? According to research in which scientists studied prehistoric earthquakes and volcanic activity, they have learned that as ice melts and waters runs off, tremendous amounts of weight are lifted off of Earth's crust. As the newly freed crust settles back to its original, pre-glacial shape, it can cause seismic plates to slip and stimulate volcanic activity. As a result, we should be seeing an   increase in both the frequency and intensity of earthquakes worldwide.

As with the Maule Earthquake which took place in Chile in February of 2010, all earthquake activity can be traced directly to the movement of one or more tectonic plates.

On February 27, 2010 a magnitude 8.8 earthquake struck Maule, Chile causing widespread damage and casualties. **The Maule earthquake ranks as one of the 10 strongest earthquakes EVER recorded!** It was also the most powerful earthquake worldwide since the 2004 Sumatran quake that triggered the massive Indian Ocean tsunami. The Maule was the

172

strongest earthquake to strike Chile since the magnitude 9.5 which hit Valdivia on May 22, 1960. A tsunami warning from the Maule quake was issued for the U.S. west coast, British Columbia, Alaska, and Hawaii.

Listed below are earthquakes worldwide ranking in the following categories: Strongest Earthquakes, Deadliest Earthquakes, and Most Destructive Earthquakes, in descending order, from 856 AD through 2010 AD.

# Earth's Strongest Earthquakes:

**May 22, 1960:** Magnitude 9.5 earthquake killed 1,655, injured 3,000, left 2,000,00 homeless, and caused $550 million damage in southern Chile; while the ensuing tsunami caused 61 deaths, $75 million damage in Hawaii; 138 deaths and $50 million damage in Japan; 32 dead and missing in the Philippines; and $500,000 damage to the west coast of the United States.

**March 27, 1964:** Magnitude 9.2 quake in Prince William Sound, Alaska, and ensuing tsunami killed 128 people (tsunami 113, earthquake 15) and caused about $311 million in property loss. Anchorage, about 120 kilometers northwest of the epicenter, sustained the most severe property damage.

## Earth's Strongest Earthquakes (cont):

**Dec. 26, 2004**: Magnitude 9.1 quake off the Indonesian island of Sumatra triggered a tsunami that killed an estimated 228,000 people in 12 countries.

**Aug. 13, 1868**: Magnitude 9.0 quake in Arica, Peru (now Chile) triggered tsunamis that killed more than 25,000 people in South America.

**Nov. 4, 1952:** Magnitude 9.0 earthquake in Kamchatka, USSR triggered a tsunami that caused property damage but no deaths in Hawaii.

**Jan. 26, 1700:** Magnitude 9.0 earthquake in the Cascadia subduction zone from mid-Vancouver Island in British Columbia along the Pacific Northwest coast.

**Nov. 25, 1833:** Magnitude 9.0 quake near Sumatra, Indonesia and ensuing tsunami caused extensive loss of life on the island.

**Feb. 27, 2010:** Magnitude 8.8 in Maule, Chile. Damage and loss of life presently unknown.

**Jan. 31, 1906:** Magnitude 8.8 off the coast of Ecuador and Colombia generated a strong tsunami that killed 500 to 1500 in coastal areas.

**Nov. 1, 1755:** Magnitude 8.7 quake in Lisbon, Portugal and ensuing tsunami killed an estimated 60,000 people and destroyed much of Lisbon.

**Feb 4, 1965:** Magnitude 8.7 quake in the Rat Islands, Alaska caused a 10.7m high tsunami on Shemya Island no loss of life.

**Jul. 8, 1730:** Magnitude 8.7 quake in Valparasio, Chile, killed at least 3,000 people.

**Mar. 28, 2005:** Magnitude 8.7 quake quake off northern Sumatra island in Indonesia.

# Earth's Deadliest Earthquakes:

**Jan. 23, 1556:** 8 magnitude quake in Shaanxi, China, 830,000 deaths.

**Jul. 27, 1976:** 7.5 magnitude quake in Tangshan, China, 255,000-655,000 deaths.

**Aug. 8, 1138:** Quake near Syria, Aleppo killed 230,000

**Dec. 26, 2004:** Magnitude 9.1 quake off Sumatra killed 227,898 people.

**Jan. 12, 2010:** Magnitude 7.0 quake near Haiti killed 222,521 people (official estimate).

# Earth's Most Destructive Earthquakes:

Abbreviated table from USGS. Listed in order of greatest number of deaths.

| Date | Location | Deaths |
|------|----------|--------|
| 1556 01 23 | Shaanxi (Shensi), China | 830,000 |
| 1976 07 27 | Tangshan, China | 255000 |
| 1138 08 09 | Syria, Aleppo | 230000 |
| 2004 12 26 | Sumatra | 227,898 |
| 2010 01 12 | Haiti region | 222,521 |

175

856 12 22 1920 1216 Iran, Damghan

Haiyuan,Ningxia (Ning-hsia), China 200,000200,000

893 03 23      Iran, Ardabil   150000


1923 09 01     Kanto (Kwanto), Japan        142

        2008 05 12     Eastern Sichuan,  China
87587 7.9

        2005 10 08     Pakistan        86000 7.6

1667 11

Caucasia, Shemakha

 80,000

1727 11 18     Iran, Tabriz    77,000

        1908 12 28     Messina

        1727 11 18     Iran             100,000

## Artistic Rendering of our Galaxy

## From Deep Space

**TWIN GALAXIES: Milky Way & Andromeda Merging to Become One...**

## Hopi Blue Kachina Dancer

# The HOPI Nation

The Hopi's created their Blue Kachina  in memory of their ancient ancestors whom have passed down a warning which reminds them to look towards the heavens for the return of the Great Blue Star ~ which will come again ~ and hail the beginning of massive planetary upheavals and Earth changes.

# 11.7

## Are You Living Next-Door to a Super Volcano?

### Increase in Volcanic Activity Worldwide

Over the past couple of years, our planet has been experiencing an unusual amount of earthquakes, volcanic eruptions and tsunamis, all of which have been showing an increasing trend in both frequency and intensity. All of these events are closely intertwined with the Earth's underlying plate structures which fit together like a giant puzzle to make up the surface of our planet.

The very thin surface, or crust, of our planet consists of 14 large separate plates, and many sub-plates. At the point of contact where two or more plates meet there is an area that is very active in both earthquake & volcanic activity. This is due to the enormous stress and friction created by the two opposing forces.

At these critical points the plate structures must slightly displace one another. Normally, the surface plate moves upwards and the other plate moves underneath it – a process called subduction. Most plate boundaries occur at the coastlines of each of our continental shelves. Needless to say, most areas where two or more tectonic plates meet are VERY unstable, resulting in a larger amount of earthquakes in that area.

According to geologists there are more than 2,000 volcanoes worldwide...and the volcanic activity on our planet is presently **stepping-it-up.**

This information may not seem important  unless you live in an area affected by volcanic activity. BUT...this information is extremely relevant to 2012 due to the fact that astrophysicists believe that volcanism is one event that will increase both in frequency and intensity as we approach our Galactic Crossing.

Within the continental United States there is only one volcano that is of major concern...and it just so happens to be one of the largest VOLCANOES on the planet! This SUPERVOLCANO is located in the famous YELLOWSTONE NATIONAL PARK.

Yellowstone contains 182 geysers and a massive underground mud volcano. The Yellowstone plateau has been **VOLCANICALLY ACTIVE for over TWO MILLION YEARS** and it has been the scene of some of our planet's largest eruptions. In addition, the Yellowstone caldera has one of the world's largest hydrothermal systems ~ making YELLOWSTONE the largest volcanic system in North America. The current caldera was formed by large eruptions 640,000 years ago and is so massive as to cover approximately 240 cubic miles in diameter. Since 2008, Yellowstone's volcanic activity has been of some concern to geologists as it has been becoming increasingly active.

Yellowstone was acting normally until 2008. Then, in late 2009 and 2010 scientists began recording what they refer to as "swarms" of earthquake activity being generated from Yellowstone...and being felt as far away as Reno, Nevada. Since then these swarms have been increasing in both frequency and intensity.

Geological records show that Yellowstone has had three major volcanic eruptions: 70,000 years ago, 160,000 years ago, and 640,000 years ago.

AND, based upon both historical data and current geothermal activity, geologists and geophysicists all believe that our super-volcano is  PAST DUE for another major eruption.

# S U P E R-V O L C A N O

## YELLOWSTONE is CURRENTLY a
## RED ALERT VOLCANO!

## Yellowstone Super-Volcano Heating Up
### 2010 Earthquake Swarms

**The January/February swarm of 2010  is now the 2nd largest swarm ever recorded at Yellowstone** ~ with over 1800 earthquakes measured.  The largest earthquake was magnitude 3.8  and  was  felt  in  the  park  and surrounding areas.  The earthquakes are believed to be tectonic in origin,  caused by regional basin and range extensions.  Some earthquakes are caused by geothermal activity.

As of April 6, 2010 a total of 2,347 earthquakes had been located for the entire swarm,  including 16 with a magnitude greater than 3.0; 141 with M2.0-2.9; 742 with M1.0-1.9; and 1,361 with M0.0-0.9.  The largest events were a pair of earthquakes of magnitude 3.7

and 3.8 that occurred after 11 PM MST on January 20, 2010. Both events were felt throughout the park and in surrounding communities in Wyoming, Montana, Idaho and Nevada.

## 2008 Earthquake Swarm (12.31.08)

"An earthquake swarm has been recorded at Yellowstone volcano. The swarm has been occurring since 26th December beneath Yellowstone Lake, three to six miles south-southeast of Fishing Bridge, Wyoming. A total of more than 250 earthquakes have been measured in the swarm. The largest earthquake was a magnitude 3.9 at 10:15 PM MST on 27th December. There have been nine earthquakes between magnitude 3 to 3.9 and around 24 between magnitude 2 to 3. Some events were felt by people in Yellowstone National Park. Yellowstone is a dormant super-volcano, and earthquake swarms, ground uplift, and geothermal activity are common. There are no other signs of unrest at the volcano."

# Smithsonian USGA
## Active & Ongoing Active
## Volcanoes Worldwide

## New Activity:

| Egon, Flores Island (Indonesia) | Etna, Sicily (Italy) | Eyjafjallajökull, Southern Iceland | Gaua, Banks Islands (SW Pacific) | Miyake-jima, Izu Islands (Japan) | Redoubt, Southwestern Alaska | Reventador, Ecuador

## Ongoing Activity:

| Arenal, Costa Rica | Batu Tara, Komba Island (Indonesia) | Dukono, Halmahera | Karymsky, Eastern Kamchatka (Russia) | Kilauea, Hawaii (USA) | Kliuchevskoi, Central Kamchatka (Russia) | Popocatépetl, México | Rabaul, New Britain | Sakura-jima, Kyushu | Shiveluch, Central Kamchatka (Russia) | Soufrière Hills, Montserrat

# UNESCO
# World Heritage
# Volcanoes

UNESCO reports and monitors all volcanoes within the United States – both active and inactive. The following is a listing volcanoes within each state:

# WORLDWIDE ERUPTIONS
## ACTIVE VOLCANOES (1979-2010)

### April 2010

**EGON Flores Island (Indonesia)** 8.67°S, 122.45°E; summit elev. 1703 m.

**ETNA Sicily (Italy)** 37.734°N, 15.004°E; summit elev. 3330 m

**EYJAFJALLAJOKULL Southern Iceland 63.63°N, 19.62°W; summit elev. 1666 m.** At 2300 on 13 April, a seismic swarm was detected below the central part of Eyjafjöll, W of the previous eruption fissures. About an hour later, the onset of seismic tremor heralded an eruption from a new vent on the S rim of the central caldera, capped by Eyjafjallajökull glacier.

The eruption was visually confirmed early in the morning on 14 April; an eruption plume rose at least 8 km above the glacier. Meltwater flowed to the N and S. News outlets reported that a circular ice-free area about 200 m in diameter was seen near the summit.

**GAUA Banks Islands (SW Pacific)** 14.27°S, 167.50°E; summit elev. 797 m

**MIYAKE-JIMA Izu Islands (Japan)** 34.079°N, 139.529°E; summit elev. 815 m

**REDOUBT Southwestern Alaska** 60.485°N, 152.742°W; summit elev. 3108 m **REVENTADOR Ecuador** 0.077°S, 77.656°W; summit elev. 3562 m **ARENAL Costa Rica** 10.463°N, 84.703°W; summit elev. 1670 m

**BATU TARA Komba Island (Indonesia)** 7.792°S, 123.579°E; summit elev. 748 m

**DUKONO Halmahera** 1.68°N, 127.88°E; summit elev. 1335 m

**KARYMSKY Eastern Kamchatka (Russia)** 54.05°N, 159.45°E; summit elev. 1536 m

**KILAUEA Hawaii (USA)** 19.421°N, 155.287°W; summit elev. 1222 m. Geologic Summary. Kilauea, one of five coalescing volcanoes that comprise the island of Hawaii, is one of the world's most active volcanoes. Eruptions at Kilauea originate primarily from the summit caldera or along one of the lengthy E and SW rift zones that extend from the caldera to the sea. About 90% of the surface of Kilauea is formed of lava flows less than about 1,100 years old; 70% of the volcano's surface is younger than

600 years. A long-term eruption from the East rift zone that began in 1983 has produced lava flows covering more than 100 sq km, destroying nearly 200 houses and adding new coastline to the island.

**KLIUCHEVSKOI Central Kamchatka (Russia)**
56.057°N, 160.638°E; summit elev. 4835 m

**POPOCATEPETL México** 19.023°N, 98.622°W; summit elev. 5426 mRABAUL

**New Britain** 4.271°S,152.203°E; summit elev. 688 m
**SAKURA-JIMA Kyushu** 31.585°N, 130.657°E; summit elev. 1117 m

Source: US Geological Survey Hawaiian Volcano Observatory (HVO) Kilauea Information from the Global Volcanism Program.

# Photo Courtesy NASA
# The Spectacular Planet Earth

This spectacular "blue marble" image is the most detailed true-color image of the entire Earth to date. Using a collection of satellite-based observations, scientists and visualizers stitched together months of observations of the land surface, oceans, sea ice, and clouds into a seamless, true-color mosaic of every square kilometer (.386 square mile) of our planet.

# 11.8

## Sleeping Beauty Finally Awakens ...But

### Is She on the Wrong Side of the Bed?

Everyone, they say, has their ups & downs, and the Sun does too! Every 11 years or so, the Sun goes from a maximum in activity to a minimum and then back to a maximum again. There are a lot of unknowns with the sun's solar cycle, but it is believed that disruptions in the Sun's magnetic field (caused by different rates of rotation within the star's polar and equatorial regions) that trigger the ebb and flow behind these changes.

189

During times of maximum activity there are lots of sunspots and solar flares as well as a small uptake in the radiation (TSI's) emitted by the Sun. Things are, of course,  vice verse for a solar minimum.

Variations in solar activity are more than a scientific curiosity.  Solar flares, for example, can disrupt or even wipe out radio communications.  Variations in TSI's can also affect global temperatures.  In fact,  it is believed that periods of extreme solar quiescence - such as the so-called Maunder Minimum - may have led to the  mini  ice age that hit regions of North America and Europe in the late 1600s.

According to Jeremy Ross, PHD (Stormx.com), solar activity is the lowest it has been for almost 100 years and may have potentially significant implications for    Earth's climate.  Sunspots are caused by the Sun's intense magnetic activity and directly impacts the solar radiation hitting our planet.

AND,  in addition  to it's normal 11-year cycle of hi and low sunspot activity, the Sun also completely reverses it's magnetic field every 22 years.  The concern is that when sunspot activity decreases only about 0.1% ~ over an extended period of time  ~ this TINY decrease can have considerable impacts on Earth's climate.  More importantly,  an increase in the Sun's radiative spectra (such as ultraviolet and cosmic rays) of more than 0.1% can affect our

190

planet's cloud cover and other aspects of our climate system.

According to research by Solanki et al, sunspot activity during recent decades has been the highest in the past 8,000 years and has likely influenced climate during the 20th century.

Over the past several years, however, the Sun has been entering another minimum in it's 11-year cycle and has been...MYSTERIOUSLY...LINGERING there! In fact, she has stayed there sooooooooo long that scientists began to wonder if the Sun would ever wake up...and cries of an impending ice age began to echo throughout the blogosphere.

The Sun continued to remain unnaturally quiet throughout 2008 and 2009, puzzling the scientific community...and worrying some scientists that the Sun may never "wake-up." Other were extremely concerned with what would occur when she did awaken. The last thing we need right now is for the Sun to "wake up on the wrong side of the bed." Then....on the 4th of January 2010...things began getting VERY interesting!

# The Sun May Be Waking up!

Suddenly...the Sun appeared to be waking up...but, perhaps she was on the **wrong side of the bed!** Not only had she been acting strangely for a couple of years...but now she was also mysteriously and literally **under attack!**

Early in 2010, the scientific community was shocked as the Sun was hit by 2 comets in January and no less than 4 comets during the month of March! And, between January 19th and January 20th of 2010 the Sun let out 5 enormous M-2 Class flares. The following is a quick summary of both the Earth and the Sun and their close encounters or highly unusual activities:

## Highly Unusual Solar Activity Begins in 2010:

-**January 4th:** Comet plunges into Sun and disintegrates upon impact.

-**January 13th:** Earth is nearly hit by asteroid 2010AL30. Approach is very close at 0.03 LD (Lunar Distance).

-**Jan. 18th – Jan. 20th**: Sun explodes with five M2 Class solar flares.

–      **Jan. 22nd** :  Comet vaporized on it's approach to the Sun.

–    **Feb. 5**th : NASA's SOHO observatory photographed an eruption of unstable magnetic loopsthat appear to have hurled some *material* in Earth's direction.

–    **Feb.8**th **– Feb.10**th : Sun crackles with more M-Class solar flares.

–    **Mar. 11**th**, 12**th**, 13**th **:** Four more comets plunge into the Sun.

–    **Mar. 16**th**:** A solar wind stream headed for Earth, along with a coronal mass ejection (CME). They created a geomagnetic storm alert for March 17th and 18th . The impacts of both events brightened Arctic skies.

It should be more than obvious that a lot of HIGH STRANGENESS has been happening in and around our planet Earth as of late? And, even the supreme "alpha" of our solar system ~ the SUN ~ has been under never before seen attacks on a enormous scale of 5 to 10 attacks per month!

Shockingly, in spite of all this the thing that most baffled and concerned scientists and astrophysicists across the planet was the fact that ~ for some unknown and NEVER BEFORE seen reason ~ the

Suns' heliosphere did not appear to be functioning properly, or perhaps not at all!

The heliosphere is a magnetized "bubble" that the Sun emits. This **bubble** covers our entire Solar System by literally protecting the Sun and all of her planets from any invading objects and galactic waves by repelling them back out into space. But now, even the Sun herself was under siege by comet and asteroid attacks on a daily basis! And, with our own planets near-hit on January 13th of 2010...it looks like we may all have a bit of a problem on our hands.

## ...BUT...

### Since I'm pretty sure this is NOT a Steven King's novel...

# WHY is this happening?

Scientists have detected a number of solar anomalies since 2008 which have led them to a possible explanation for what is going on. They believe that when the Sun took so long to close her last solar cycle and begin the new one ~ she somehow failed to close her heliosphere completely ~ as she ALWAYS does between solar cycles.

What effects will this have on our planet and the Solar System as a whole? No one really knows...we

will all just have to wait and see. But ~ one thing is
for sure ~ this leaves the **front door WIDE open,** not
only for the Sun, but for the Earth and all of the
planets in our Solar System ...leaving all of us
vulnerable to comet and asteroid attacks, as well as
to a variety of cosmic ray bursts...and highly charged
galactic waves. Most galactic waves are completely
invisible to the naked eye, but some can dance
and play and create magnificently, mysterious
displays within the night skies.

Another mysterious part of all of this is the fact that
many ancient cultures ~ one being the Hopi Indians of
Arizona ~ have native legends that talk about having
seen the very mysterious sights that are happening
now in our skies. The Hopis called them the "blue
light bursts" from the sky. This is just one more clue
that whatever is going on TODAY has happened
before within the ancient past, and perhaps MANY
TIMES within our ancient past!

# The Sun & Our Star System

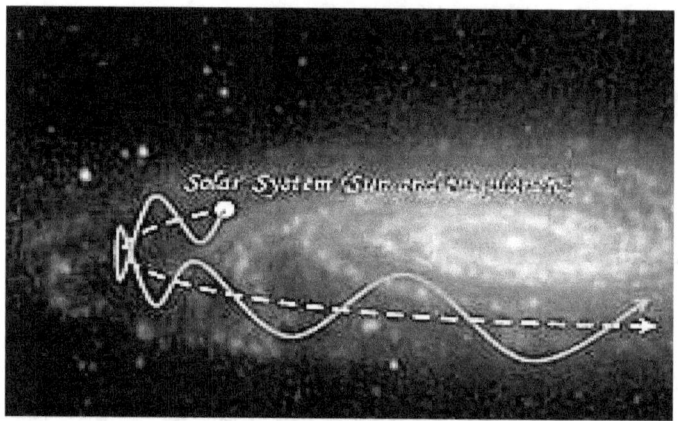

## In our Galactic Orbital Path

## Infrared View of our Galactic Center

Some legends even point out that these cosmic bursts are coming from the area of the Dark Rift or Galactic Center. In fact, the Hopi Indians have even dedicated a Kachina specifically to commemorate this event, and named him **The Blue Kachina Star.** The ancient Mayans also noted these cosmic ray events by creating a Mayan deity in it's honor named **Hunab Ku**. Literally translated, Hunab Ku means **"The Supreme God,"** however, the Maya also refer to Hunab Ku as **"The Great Architect of the Universe,"** and depicted their deity with a number of symbolic drawings.

Is this uptake in solar flares an anomaly ~ or a sign that we are headed for a more dynamic and energetic sunspot cycle than what has been predicted? This is a unique event to the scientific community. As a result no one really knows...we will all just have to wait and see. However, one thing is for certain...it leaves the **front door WIDE open**...not only for the Sun, but for the Earth and all of the planets in our Solar System, leaving all of us vulnerable to comet and asteroid attacks, as well as to a variety of cosmic ray bursts and highly-charged galactic waves. Most galactic waves are completely invisible to the naked eye, but some can dance and play and create magnificent, mysterious displays within the night skies.

## OTHER New & Strange SOLAR Discoveries:

### Oct. 30, 2008:  Magnetic Portals Connect Sun and Earth Discovered

http://science.nasa.gov/headlines/y2008/30oct_ftes.htm?list91147

**NASA:** "During Flux Transfer Events (FTE) the Earth's magnetic field presses against the Sun's magnetic field. Approximately every 8 minutes, the two fields briefly merge or "reconnect," forming a portal through which particles can flow. These magnetic cylindrical portals tend to form above Earth's equator and then roll over Earth's winter pole. In December, FTEs roll over the north pole; and in July they roll over the south pole."

### Nov. 19, 2008: NASA Discovers Cosmic Rays from a Mysterious, Nearby Object

**NASA:** "An international team of researchers has discovered a puzzling surplus of high-energy electrons bombarding Earth from space. The source of these cosmic rays is unknown, but it must be close to the solar system and it could be made of dark matter. http://science.nasa. gov/headlines/2008/19nov_cosmicrays.htm?list911479

### Dec. 6, 2008:  Two Cosmic Ray Hot Spots Never Before Seen

**MILAGRO OBSERVATORY:** An international team of researchers, using Los Alamos National Laboratory's Milagro Observatory has discovered, for the first time, 2 distinct *hot spots* (red regions) that appear to be bombarding Earth with an excess of cosmic rays. The hot spots are located near the Orion constellation.

## Dec. 26, 2009:  Earth Close to Unexpected Interstellar Cloud

**NASA:** "The fluff is held at bay just beyond the edge of the solar system by the sun's magnetic field, which is inflated by solar wind into a magnetic bubble more than 10 billion kilometers wide. Called the heliosphere, this bubble acts as a shield that helps protect the inner solar system from galactic cosmic rays and interstellar clouds. NASA's two Voyagers are located in the outermost layer of the heliosphere, where the solar wind is slowed by the pressure of interstellar gas. The fact that the fluff is strongly magnetized means that other clouds in the galactic neighborhood could be too. Eventually, our solar system will run into some of this fluff, and its strong magnetic fields could compress our heliosphere even more than it is compressed now. Additional compression could allow more cosmic rays to reach the inner solar system, possibly affecting terrestrial climate and the ability of astronauts to travel safely through space."

## Jan. 20, 2010: What is the DARK DISK Moving in Front of Epsilon Aurigae?

**Astronomy Magazine:** In August 2009, professional and amateur astronomers reported that the bright star Epsilon Aurigae had begun to lose brightness for the first time in 27 years. Epsilon Aurigae is a mysterious, bright, eclipsing, binary star that has puzzled astronomers for 175 years, and, under normal circumstances is bright enough to be seen from even the brightest of cities with only the naked eye.

199

## Feb.20th-22nd2010

## NASA: The Sun Develops 1-Million-Km Long TAIL?

"Spacecraft and amateur astronomers alike are monitoring a staggeringly-long filament of magnetism on the sun. It stretches more than one million kilometers around the sun's southeastern limb. The filament remained mostly stable for at least 2 days. However, similar filaments in the past have been known to collapse, and when they hit the surface of the sun a tremendous explosion called a "hyderflare" results. Hyderflares sometimes rival the strongest flares produced by sunspots. Solar physicists have not yet learned to predict hyderflares."

## Sun Blamed for Warming of Earth & Other Planets:

The Earth has been heating up lately, but so has Mars, Pluto and all of the other planets in our Solar System. This had lead some scientists to speculate that the recent changes in our Sun's activity has, most likely, the common thread linking all these baking events.

# Other Related:

## Cosmic, Gamma Ray Bursts:
http://news.nationalgeographic.com/news/2009/04/09
0403-gamma-ray-extinction.html

## Galactic Superwaves:

Galactic Superwaves (mysteryofthegods.blogspot.com)
http://mysteryofthegods.blogspot.com/2010/01/galacti
c-superwaves.html

## Superwave: Project Camelot & interviews Dr Paul LaViolette
http://www.youtube.com/watch?v=oURVtGKW420

## Superwave Theory: Predictions and their Subsequent Verification

http://www.etheric.com/LaViolette/Predict.html

## Exopolitics: Has the Galactic Superwave of 2012 Begun?

http://exopolitics.blogs.com/exopolitics/2005/03/did a
neutrons.html

## The Mystery of CROP CIRCLES

Mysterious Crop Circles have been appearing everywhere on the planet on a regular basis for decades ~ but have increased in frequency and complexity within the past few years.

# 11.9

## NASA's
## Official
## 2012 Prediction
## "The Day After Tomorrow"

### NASA's OFFICIAL 2012 Prediction:

"There will be a deterioration of Earth's magnetosphere during Sun's polar shift in 2012. Scientists have found two large leaks in Earth's magnetosphere, the region around our planet that shields us from severe solar storms."

### ... ...And...Just to be certain......

NASA sent up at least two more rockets within the past two years to further observe these changes!

In 2006, NASA published a very important paper about our planet's magnetosphere, and indicated a movie, which is **very, very rare for NASA to do!** The name of the movie was... **THE DAY AFTER TOMORROW!** To read this publication search GOOGLE with the key words – NASA Solar Flares 2012.

## As reported by NASA Scientists:

"Scientists have now identified two areas of concern with respect to our upcoming Galactic Alignment. Of MAJOR concern will be the deterioration of Earth's magnetosphere during the Sun's 2012 Polar Shift."

# ...Here's why...

**NASA:** "Scientists have found two large leaks in Earth's magnetosphere ~ the region around our planet that shields us from severe solar storms. In 2012 the Sun's poles will reverse, during which time our planet will experience a massive solar storm." "Usually this is no problem, but now Earth's protective magnetosphere may fail us due to the fact that it is cracked; thereby allowing violent solar and electromagnetic radiation to make it through to the surface of our planet...and causing many problems to life as we know it (eg: disabling communication satellites, mobile phones, effective sleep patterns, & causing the radiation poisoning of humans). Also, as Earth absorbs extra radiation and energy it will most likely cause changes within our planet's core, forming new volcanoes and crustal movement."

## 16 December 2008Author:

## Credit: Science@NASA

### NASA's THEMIS Spacecraft has
**discovered a breach in Earth's magnetic field 10 TIMES LARGER than anything previously thought to exist.** Solar Wind can flow in through this opening and load up the magneto-sphere, creating powerful geomagnetic storms. But, the breach itself is not the biggest surprise. Researchers are even more amazed at the strange and unexpected way it forms, overturning long-held ideas of space physics.

"At first I didn't believe it," stated David Siebeck, a project scientist from the Goddard Space Flight Center. "This finding fundamentally alters our understanding of the solar wind-magnetosphere interaction."

The magnetosphere is a bubble of magnetism that surrounds Earth and protects us from the solar wind. Exploring the bubble is a key goal of the THEMIS mission, launched in February 2007. The big discovery came on June 3,2007, when our five space probes serendipitously flew through the breach just as it was opening. On-board sensors recorded a torrent of solar wind particles streaming into the magnetosphere, signaling an event of unexpected size and importance.

**"The opening was huge—4 times wider than the Earth itself,"** says Wenhui Li, a space physicist at the University of New Hampshire who has been analyzing the data. Li's colleague Jimmy Raeder, also of New Hampshire, says "1027 particles per second were flowing into the magnetosphere—that's **a 1 followed by 27 zeros.** This kind of influx is an order of magnitude greater than what we thought was possible."

The event began with little warning when a gentle gust of solar wind delivered a bundle of magnetic fields from the Sun to Earth. Like an octopus wrapping its tentacles around a big clam, solar magnetic fields draped themselves around the magnetosphere and cracked it open. The cracking was accomplished by means of a process called "magnetic reconnection." High above Earth's poles, solar and terrestrial magnetic fields linked up (reconnected) to form conduits for solar wind. Conduits over the Arctic and Antarctic quickly expanded – and, within minutes they overlapped over Earth's equator to create the biggest magnetic breach ever recorded by an Earth-orbiting spacecraft.

The shear size of the breach is what took researchers by surprise. **"We've seen things like this before,"** says Raeder, **"but never on such a large scale. The entire day-side of the magneto-sphere was open to the solar wind."**

The circumstances were even more surprising. Space physicists have long believed that holes in Earth's magnetosphere open only in response to solar magnetic fields that point south.  The great breach of June 2007, however,  opened in response to a solar magnetic field that pointed north.

"To the lay person, this may sound like a quibble, but to a space physicist, it is almost seismic," says Sibeck. "When I tell my colleagues,  most react with  skepticism, as if I'm trying to convince them that the sun rises in the west."

## Why Scientists Can't Believe Their Eyes!

The solar wind presses against Earth's magnetosphere almost directly above the equator where our planet's magnetic field points north. Suppose a bundle of solar magnetism comes along, and it points north, too. The two fields should reinforce one another, strengthening Earth's magnetic defenses and slamming the door shut on the solar wind.  In the language of space physics, a north-pointing solar magnetic field is called a "northern IMF" and it is synonymous with **shields up!**

"So, you can imagine our surprise when a northern IMF came along and shields went down instead," says Sibeck. "This completely overturns our understanding of things." "Northern IMF events don't actually trigger geomagnetic storms," notes Raeder, " but they do set the stage for storms by loading the magnetosphere with plasma." A loaded magneto-sphere is primed for auroras, power outages, and other disturbances that can result when, say, a CME (coronal mass ejection) hits. The years ahead could be especially lively!

## Raeder explains:

"We're entering Solar Cycle 24. For reasons not fully understood, CMEs in even-numbered solar cycles (like 24) tend to hit Earth with a leading edge that is magnetized north. Such a CME should open a breach and load the magnetosphere with plasma just before the storm gets underway. It's the perfect sequence for a really big event." Sibeck agrees. "This could result in stronger geomagnetic storms than we have seen in many years."

**Author: Dr. Tony Phillips | Credit: Science@NASA**

# 11.10

# Earth
# Under
# Siege

## Near Earth "Hits" Increase 20x's since 1995

According to the most current data released by NASA there have been 7,036 near-Earth objects discovered as of May 11, 2010.  Of these, 807 are asteroids with a diameter of 1 kilometer or larger, and 1,127 are classified as PHAs ~ Potentially Hazardous Asteroids.  NASA defines a PHA as one that has a MOID* of less than 0.05 AU,** and an absolute magnitude (H) of 22.0 or less.  In addition, it is important to take the objects LD*** Miss Distance into account...as this number brings the object extremely close to the Earth.

An object that is listed as a PHA does not mean that it WILL impact Earth, but only that it shows the POTENTIAL to become a THREAT to our planet by making a close approach.  NASA then monitors these objects and updates  their  orbits in order to better predict  which objects will become an Earth-impact threat.

Their data,  which follows,  shows a definite trend in the number of Near Earth Asteroids around our planet ...and it is also clear that **the trend has been INCREASING EXPONENTIALLY by 20x's since 1995.**

# Near Earth Asteroids:

-     1995    335 NEAs
-     1999    880 NEAs
-     2000    1243 NEAs
-     2004    3144 NEAs
-     2009    6647 NEAs

       *MOID = Minimum Orbit Intersection Distance

**AU = 1 AU is equal to approximately ~150 million kilometers

***LD = Lunar Distance is equal to approximately ~ 384,000 kilometers

The following is a summary of NASAs "Recent Close Approaches to Earth" tables for the past several months. Type I Asteroids are those that meet NASAs criteria for a Potentially Hazardous Asteroid, and Type II Asteroids are those that are "on the fringe" of NASAs criteria – but of concern because of their enormous size or proximity.

# RECENT CLOSE APPROACHES
# To EARTH

|  | Total PSAs | TypeI  PSAs |
|---|---|---|
| **May 2010** | 19 | 7 |
| **June 2010** | 18 | 3 |
| **July 2010** | 18 | 2 |
| **August 2010** | 18 | 3 |

In conclusion, NASAs official data shows a direct increase in the number of comets and asteroids in and around our planet as we approach the Galactic Center or Dark Rift in 2012. This is "AS EXPECTED" since the Galactic Center Region is known to be littered with so much space "debris" that it appears dark and clouded on a regular basis.

NASA's data very clearly shows that there has been a DRAMATIC and exponentially ESCALATING number of Near Earth Objects since 1995. **And...by**

**2009...this number has increased 20 fold from 335 to 6,647.** As of the date of this writing we have approximately 18 Near Earth Objects within the vicinity of our planet at any given time, and an average of 3 of these NEAs are considered to be Potentially Hazardous Asteroids.

Once again, these numbers coincide with what is expected as we approach our Galactic Crossing on 12:21:12...and these numbers should continue on an ESCALATING TREND though that date. There is also the added factor that our Sun has failed to close her heliosphere...allowing all objects that would normally be repelled back into space to very easily enter into our Solar System. It also appears that the Sun has some type of "pull" or ability to attract these objects to itself...as seen by the enormous number of solar impacts within the first couple of months of 2010. This should be a positive thing for our planet as it will allow them to avoid the Earth as a point of collision.

Oddly enough it appears that our SUN is doing all she can to protect planet Earth and the rest of her entourage ...and...as long as an object is not large enough to disrupt the Sun into some sort of "massive flare" event we should be able to make it through this portion of the 2012 phenomenon.

A large rocky body in orbit about the Sun is referred to as an asteroid whereas much smaller particles in orbit about the Sun are referred to as meteoroids.

Once a meteoroid enters the Earth's atmosphere and vaporizes, it becomes a shooting star.  But, if a meteoroid survives its fiery passage through the Earth's atmosphere and lands upon the Earth's surface it is called a meteorite.  Many comets generate meteoroid streams when their icy cometary nuclei pass near the Sun and release the dust particles that were once embedded in the cometary ices. These meteoroid particles then follow in the wake of the parent comet and are seen as what we describe as the "tail" of a shooting star.

### How Many Near-Earth Objects Have Been Discovered so Far?

**As of May 11, 2010:**  7036 Near-Earth objects have been discovered.  807 of these NEOs are asteroids with  a diameter of approximately 1 kilometer or larger.  1119 of these NEOs have been classified as Potentially Hazardous Asteroids (PHAs)

### WHAT IS A PHA?

**P**otentially **H**azardous **A**steroids (PHAs) are currently defined based on parameters that measure the asteroid's potential to make threatening close approaches to the Earth. PHAs are all asteroids that meet the following criteria: those that have an Earth Minimum Orbit Intersection Distance of less than and up to 0.05 AU. - those that have an absolute magnitude (H) of 22.0 or less.

So, according to NASA , all asteroids with an Earth Minimum Orbit Intersection Distance (MOID) of 0.05 AU or less and an absolute magnitude (H) of 22.0 or less are considered PHAs.  There are currently 1127 known PHAs.

This potential to make close Earth approaches does **not** mean a PHA **will** impact the Earth. It only means there is a possibility for such a threat. By monitoring these PHAs and updating their orbits as new observations become available,  we can better predict the close-approach statistics and thus their Earth-impact threat.

**Notes: LD means "Lunar Distance." 1 LD = 384,401 km, the distance between Earth and the Moon. 1 LD also equals 0.00256 AU. MAG is the visual magnitude of the asteroid on the date of closest approach.**

# 11.11

## Galactic SuperWaves Found Layered Into Earth's Ancient Past

**Dr. Paul LaViolette ~ Discovers Cosmic Dust & Galactic Superwaves Layered into Earth's History** Dr. Paul LaViolette holds 9 degrees in physics from Johns Hopkins University, an MBA from the University of Chicago, and a PhD from Portland State University.

Dr. LaViolette is the credited for being the scientist who first discovered Cosmic Dust within ancient ice core samples. Recently, he has also discovered evidence of the existence and bombardment of our planet by Galactic Superwaves!

Dr. LaViolette has published many original papers in physics, astronomy, climatology, systems theory, and psychology; and, has served as a Solar Energy Consultant for the United Nations, the Greek government, and The Club of Rome's "Goals for Mankind" Project. He has also consulted numerous Fortune 500 companies on new ways of stimulating innovations.

## About Dr. LaViolette: Recognized in Marquis "Who's Who in Science and Engineering," Dr.

LaViolette was the first to predict that high intensity "volleys" of cosmic ray particles travel directly to our planet from distant sources in our Galaxy, a phenomenon now confirmed by scientific data. He was also the first to discover high concentrations of Cosmic Dust in our Ice Age polar sheets - indicating the occurrence of a global cosmic catastrophe in ancient times. Based on his work, he made predictions about the entry of interstellar dust into our solar system ten years before its confirmation in 1993 by data from the Ulysses Spacecraft and by radar observations from New Zealand.

Dr. LaViolette is currently president of The Starburst Foundation's "Interdisciplinary Scientific Research Institute." One key area of Starburst's research is concerned with the investigation of Galactic SuperWaves ~ intense cosmic ray particle barrages that travel to us from the center of the Milky

Way Galaxy. and last for periods of up to a few thousand years.

One thing which is relevant for the 2012-2013 Nexus Event is that the Starburst Foundation has discovered that at least TWO Superwaves with the strength to generate new Ice Ages that are traveling our way from their place of origin around 26,000 light years away! This distance appears to place their origin at the center of the Milky Way Galaxy...the Galactic Center...or...the Dark Rift!.

Dr. LaViolette began alerting the scientific community to the existence of SuperWaves in 1983 through published papers and scientific conference presentations. He also raised public awareness about the Superwave phenomenon through his book Earth Under Fire as well as through various magazine articles.

Many aspects of Dr. LaViolette's Superwave Theory have since been verified by recent observations. The primary goal of Dr. LaViolette's research has to do with GALACTIC CORE EXPLOSIONS ~ violent explosions occurring within the centre regions of galaxies. His research during early 1980's led him to the ice core samples from Greenland and Antarctica ~ and it was, particularly in Greenland, that he found high levels of Cosmic Dust within the Ice Age portion of the ice-core samples. This also confirmed his hypothesis that there would been an arrival of Cosmic Rays from the Galactic Centre around that time.

According to Dr. LaViolette, "The arrival of these Cosmic Rays pushed Galactic Dust into our Solar System, causing extreme climatic change on all of the planets within our Solar System. The Cosmic Rays were the cause behind our Solar System filling up with Cosmic Dust...and covering the Sun so that the radiation that reached the Earth began coming in at a different spectrum."

In a physical view, the doctor is speaking about a more "reddish and dusky sky" which created a difficulty in being able to see the stars in the nighttime sky. But the major problem was that the radiation that reached the Earth was within the infrared spectrum...creating what Dr. LaViolette calls an "interplanetary hot house effect."

## Our Heliosphere

Like Earth, our entire Solar System has its own atmosphere, called the heliosphere. This "bubble" surrounds the Sun and her planets as they travel through galactic space. Like our Earth's magnetosphere, the movement of the heliosphere creates a rounded "head" and a narrowing "tail"~ making it more of an egg shape.

Until recently, astronomers believed that our Solar System was a region relatively free from Cosmic Dust. They believed that the Cosmic Dust and frozen material from space were kept outside our protective bubble or heliosphere. This was confirmed when the IRAS and Ulysses Spacecrafts showed infrared images of the Solar System, surrounded by wispy clouds of Cosmic Dust that increases in density just beyond Saturn.

# ...So...

If the Cosmic Dust is surrounding the heliosphere, what would make it suddenly able to enter into the heliosphere, and how would this coincide with huge Solar Flares? Dr. LaViolette envisioned something disrupting the heliosphere from the outside – impacting it – and drawing Cosmic Dust inside along with it, thereby energizing the Sun. The energy of such an impact would have to be immense, and the most logical place to look for such an ENORMOUS energy force would have to be the center region of the Milky Way Galaxy. Perhaps related to this is the puzzling fact that, even though we have not witnessed any Galactic Explosions or "bursts", the measurements of Cosmic Dust streaming inside the heliosphere has been steadily increasing ~ by almost 3X's ~ since our last solar maximum in 2001.

During the solar maximum of each 11 year cycle, the polarity of the Sun shifts - North becomes South and visa versa. This brief period of magnetic instability allows some Cosmic Dust to enter the heliosphere because the Sun's "shields" are reduced. But once the new polarity is established, the Sun is usually extremely quick to block the dust. **THIS TIME IT DID NOT HAPPEN!** Cosmic Dust has been steadily streaming in from the Galactic Center and astronomers are at a loss to explain why.

This is the first time something like this has happened, at least within the time frame of our solar monitoring. Not to mention so many anomalies and newly discovered phenomenon that have appeared this year, together with the dramatic climate changes currently taking place on all of the planets within our Solar System. And, these changes are far from over. It's likely that our Solar System is already experiencing the invasive energy from the Galactic Equator as we move into position to align with it in 2012. To makes things appear even worse, recent data shows that dramatic and potentially deadly effects can result from solar flares and coronal mass ejections.

Substantial data suggests that an event, similar to the one anticipated in the 2012 "doomsday" scenario, occurred about 14,950 years ago and was recorded by ancient humans. This event appears to have lasted for several years in duration, and was responsible for the abrupt end of the last Ice Age as well as a substantial "culling" of the human population.

# EXTREME SOLAR EVENT in 2012

The surprising findings of Dr. LaViolette, supported by other research, suggests that this extreme solar event corresponded to powerful radiation coming from the center of the Milky Way Galaxy, and was associated with Gamma Rays and Cosmic Dust. Recent observations have shown a dramatic increase in gamma ray energy within the Galaxy's equator region. In addition, these gamma rays are expected to be at their maximum during the alignment with our Solar System on 21 December 2012.

Past Ice Core records (strata from 13,880 to 13,785 BCE) also suggest that intense radiation from this last event could have lasted many years. According to Dr. LaViolette, **it seems highly likely that this alignment will cause another EXTREME SOLAR EVENT since other factors precipitating a "solar maximum" (ie: the opposition of major planetary barycenters) also converge on this exact date.**

Based upon his theories, Dr. LaViolette believes that since all galactic centers routinely radiate lethal gamma rays, it unlikely that life ~ at least as we understand it ~ can survive in the Universe. Sooner or later it is destined to be zapped!

I could not leave all of you with such horrific news...there has to be a brighter side to this story...

.      ...<u>AND THERE IS!</u>

As being ONE who has the wisdom and courage

to seek out the Knowledge and the Truth

...which you are since you have this book in hand...–

there IS some GOOD NEWS that must be

destined specifically for YOU!

...AND...FINALLY...

# NOW for the GOOD NEWS!

...A new genetic study of "Y-chromosome" variations by Dr. Marcus Feldman of Stanford University shows that the population from which the world's present-day population is derived consisted of only about 2,000 individuals. What Dr. Feldman is saying is that at the time of our planet's last extinction 2,000 humans DID SURVIVE! And, that should be exhilarating news for all of us!

## ...SO...

## Somehow...Humans, Flora & Fauna

# DID SURVIVE

## Past Doomsday Events!

# ...PROVING THAT...

At the Very Least ~
Some  of  Us
WILL   SURVIVE
2 0 1 2!

# 11.12

## Giant SuperWaves Target Earth

Brighter Than Sun Emissions Over 150,000 Years

Brightest Explosion Ever Recorded

**from RedNovaNews Website**

**2005/02/18 NASA:** Scientists have detected a flash of light from across the Galaxy so powerful that it bounced off the Moon and lit up the Earth's upper atmosphere. The flash was brighter than anything ever **detected from beyond our Solar System and lasted over a tenth of a second**.

NASA and European satellites and many radio telescopes detected the flash and its aftermath. On December 27, 2004, two science teams reported on this event at a special press conference at NASA headquarters. In addition, a multitude of papers were planned for publication. The scientists said the light came from a "giant flare" on the surface of an exotic neutron star, called a **magnetar**. The apparent magnitude was brighter than a full moon and all known, historical star explosions. The light was brightest in the gamma-ray energy range, far more energetic than visible light or X-rays and invisible to our eyes. Such a close and powerful eruption raises the question of whether an even larger influx of gamma rays, disturbing the atmosphere, was responsible for one of the mass extinctions known to have occurred on Earth hundreds of millions of years ago.

Also, if giant flares can be this powerful, then some gamma-ray bursts (thought to be very distant black-hole-forming star explosions) could actually be from neutron star eruptions in nearby galaxies. NASA's newly launched Swift Satellite and the NSF-funded Very Large Array (VLA) were two of many observatories that observed the event, arising from neutron star SGR 1806-20, about 50,000 light years from Earth in the constellation Sagittarius. "This might be a once-in-a-lifetime event for astronomers, as well as for the neutron star," said Dr. David Palmer of Los Alamos National Laboratory. Dr. Palmer is the lead author on a paper describing the Swift observations. "We know of only two other giant flares in the past 35 years, and this December event was one hundred times more powerful."

WHAT IS A NEUTRON STAR? A neutron star is the core remains of a star once several times more massive than our Sun. When such stars deplete their nuclear fuel, they explode ~ an event called a supernova. The remaining core is dense, fast-spinning, highly magnetic, and only about 15 miles in diameter. Millions of neutron stars fill our Milky Way Galaxy at any given time.

Scientists have discovered about a dozen ultra high-magnetic neutron stars, called magnetars. The magnetic field around a magnetar is about 1,000 trillion gauss, strong enough to strip information from a credit card at a distance halfway to the moon. (Ordinary neutron stars measure a mere trillion gauss; the Earth's magnetic field is about 0.5 gauss.)

Four of these magnetars are also called soft gamma repeaters, or **SGRs**, because they flare up randomly and release gamma rays. Such episodes

release about 10^30 to 10^35 watts for about a
second, or up to millions of times more energy than
our Sun. For a tenth of a second, the giant flare on
SGR 1806-20 unleashed energy at a rate of about
10^40 watts.

**The total energy produced was more than
the Sun emits in 150,000 years!** "The next
biggest flare ever seen from any soft gamma
repeater was peanuts compared to this incredible
December 27 event," said Gaensler. "Had this
happened within 10 light years of us, it would have
severely damaged our atmosphere. Fortunately, all
the magnetars we know of are much farther away
than this."

    A scientific debate raged in the 1980s over whether
gamma-ray bursts were star explosions from beyond
our Galaxy or eruptions on nearby neutron stars. By
the late 1990s it became clear that gamma-ray bursts
did indeed originate very far away and that SGRs
were a different phenomenon. But the extraordinary
giant flare on SGR 1806-20 reopens the debate,
according to **Dr. Chryssa Kouveliotou** of NASA
Marshall Space Flight Center, who took part in both
the Swift and VLA analysis."A sizable percentage of
"short" gamma-ray bursts, less than two seconds,
could be SGR flares," she said. "These would come
from galaxies within about a 100 million light years
from Earth. Long gamma-ray bursts appear to be
black hole forming explosions many light years away.

# 11.13

# Cosmic Rays Hit Space Age High

**Richard Mewaldt**  (Caltech) & **Dr. Tony Phillips**
(Heliophysics News Team          09.28.09

**NASA's** Advanced Composition Explorer
Spacecraft (ACE) revealed that cosmic ray levels
have jumped 19% above the previous Space Age
high.

"In 2009, cosmic ray intensities have increased 19%
beyond anything we've seen in the past 50 years,"
says Richard Mewaldt of Caltech.  "The increase is
significant, and it could mean we need to re-think how
much radiation shielding astronauts take with them on
deep-space missions."

The cause of the surge is solar minimum, a deep lull in solar activity that began around 2007 and continues today.  Researchers have long-known that cosmic rays go up when solar activity goes down. Right now  solar activity is as weak as it has been in modern times, setting the stage for what Mewaldt calls   "a perfect storm of cosmic rays."

**"We're experiencing the deepest solar minimum in nearly a century," says Dean Pesnell of the Goddard Space Flight Center, "so it is no surprise that cosmic rays are at record levels for the Space Age."**

Galactic cosmic rays come from outside the solar system.  They are subatomic particles, mainly protons, but some have been nuclei-accelerated to almost  the speed of light by distant supernova explosions.  Cosmic  rays  cause air showers of secondary particles when they hit the Earth's atmosphere. They also pose a health hazard to astronauts and a single cosmic ray  can  easily disable a satellite if  it  hits  an  unlucky integrated circuit.

The sun's magnetic field is our first line of defense against these highly-charged, energetic particles. Our entire solar system from Mercury to Pluto and beyond is surrounded by a bubble of solar magnetism called "the heliosphere."  It springs from the sun's inner magnetic dynamo and is inflated to gargantuan proportions by the solar wind. When a cosmic ray tries to enter our solar system, it must fight through our heliosphere's outer layers. Then, if it is able to make it's way past that barrier, there is a thicket of magnetic fields waiting to scatter and deflect the intruder. "The heliospheric sheet is shaped like a ballerina's skirt," according to J. R. Jokipii of The University of Arizona. "At times of low solar activity, this natural shielding is weakened... allowing more cosmic rays to reach the inner solar system," explains Pesnell.

Mewaldt lists three aspects of the current solar minimum that are combining to create the perfect storm:

1)The sun's magnetic field is weak. "There has been a sharp decline in the sun's interplanetary magnetic field (IMF) down to only 4 nanoTesla (nT) from typical values of 6 to 8 nT," he says. "This record-low IMF undoubtedly contributes to the record-high cosmic ray fluxes."

2)The solar wind is flagging. "Measurements by the Ulysses spacecraft show that solar wind pressure is at a 50-year low," he continues, "so the magnetic bubble that protects our solar system is not being inflated as much as usual." A smaller bubble gives cosmic rays a shorter-shot into the solar system. Once a cosmic ray enters the solar system, it must "swim upstream" against the solar wind. Solar wind speeds have dropped to very low levels in 2008 and

2009, making it easier than usual for a cosmic ray to proceed.

3)The sun's current sheet is flattening. Imagine the sun wearing a ballerina's skirt as wide as the entire solar system with an electrical current flowing along the wavy folds. That is the "heliospheric current sheet," a vast transition zone where the polarity of the sun's magnetic field changes from plus (north) to minus (south). The current sheet is important because cosmic rays tend to be guided by its folds. Lately, the current sheet has been flattening itself out, allowing cosmic rays more direct access to the inner solar system.

**"If the flattening continues as it has in previous solar minimums, we could see cosmic ray fluxes jump all the way to 30% above previous Space Age highs," predicts Mewaldt.**

He goes on to explain that the Earth is not in any great peril from these extra cosmic rays. Our planet's atmosphere and magnetic field combine to form a formidable shield against space radiation, protecting all of us on the surface. Indeed, we've weathered storms much worse than this. Hundreds of years ago, cosmic ray fluxes were at least 200% higher than they are now.

232

Leading scientific researchers know this  because when cosmic rays hit the atmosphere,  they produce an isotope of beryllium (10Be) which is preserved within our polar ice caps.  By examining ice core samples,  it is possible to estimate cosmic ray fluxes more than a thousand years into the past.  Even  with the  recent surge,  cosmic rays today  are  much weaker  than they have been at various times in the past millenniums.


**"The space era has so far experienced a time of relatively low cosmic ray activity," says Mewaldt. "However, we could now be returning to levels typical of past centuries."**

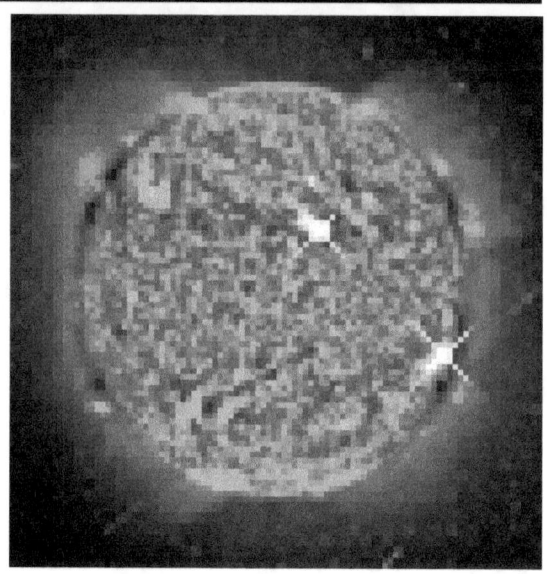

# Rare & Stunning
# Photo of the Sun

# XII

## And...
## What About our
## Beloved
## Goddess
## The Sun?

Just like so many of us....the Sun also has her
ups and downs.  Every 11 years (or so) the Sun goes
through a cycle ranging from maximum to minimum
solar activity. Scientists  study  and categorize  this
activity by the amount and intensity of her solar flares.
Flares are categorized according to the strength of
they emit – ranging from the lowest class A flare to B,

C, M~ and finally the strongest of all Solar Flares~the X flare.

The Sun completed Cycle 23 in 2008. Cycle 23 had peaked in 2000, and then remained in solar minimum throughout 2008. In fact, she stayed there SOOOOO long that scientists were beginning to wonder if our beloved Sun would ever wake up! Finally, in December of 2008, the Sun awoke and began Cycle 24.

Cycle 24 is predicted to be a very weak cycle. Cycle 23 had an average sunspot number of 120, whereas Cycle 24 is expected to reach a maximum of only around 90, peaking in 2013. But, just like all of her planets, the Sun herself has been acting a **bit out of character.**

In January of 2010...our long-quiet Sun became suddenly active, and startled scientists by emitting a large number of solar flares. On January 17th and 18th the Soviet TESIS Satellite observed moderate Class C flares, and then on January 19th there were a series of increasingly intense flares that culminated with two Class M flares! These were the largest flares seen since the summer of 2007.

Basically, the Sun went from her lowest activity in 100 years ~ and awakened to a robust morning ~ displaying not only a LARGE number of flares, but flares of increasingly strong intensity, ending with two

M class flares! Is this merely a solar anomaly, or a sign that we a headed for a much more dynamic and energetic cycle than predicted? Scientists continue to watch and evaluate...and have no idea where our Sun is  headed ~ according to them ~ only time will tell!

# Our Solar System is in Hyper-Dimensional Change

Physicists, astrophysicists and other scientists agree that our entire solar system is currently entering into a much  more  highly  electrically-charged  part   of  our galaxy.  In fact, the galactic energy is increasing to such an extreme extent as to be considered **HYPER-DIMENSIONAL in nature!**

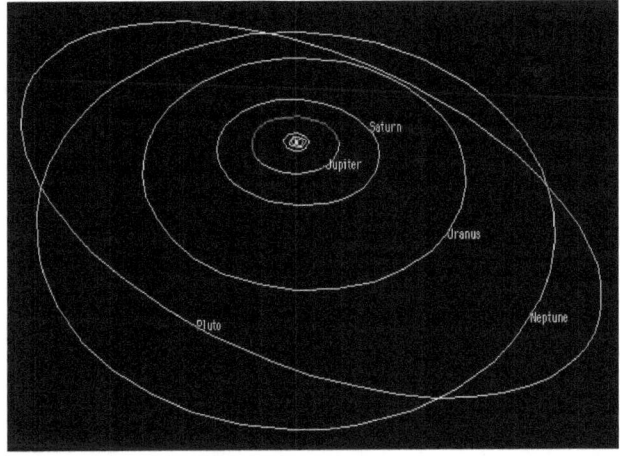

**PLANETARY ORBITS around the SUN**

**Beginning at the planet furthest from the Sun
these changes are summarized as follows:**

# PLUTO

Like our planet, Pluto is also experiencing global
warming, but ~ in addition ~ Pluto has had a dramatic
increase in atmospheric pressure by 300! Pluto has
also shown dramatic color change, having  become
significantly redder and brighter.

# NEPTUNE

Neptune was once known for her **Great Red Spot** -~
which was the size of about ½ the diameter of the
Earth.  However, Neptune's Great Red Spot has
recently disappeared.  The planet is also showing a
significant and progressive increase in brightness and
cloudiness. Clouds began increasing significantly in
2002.

# URANUS

Uranus has recently developed a large vortex...huge
enough to engulf 2/3 of the United States. And, like
Pluto and Neptune, the planet is also experiencing an
overall increase in atmospheric activity.

## SATURN

Saturn now has an ENORMOUS new ring that extends 170 million miles downwards, and is equivalent to 20 times her diameter! It would take about one billion Earths stacked together to fill Saturn's new ring.

## JUPITER

Jupiter is in the midst of VIOLENT global climate change. Violent masses of storms with winds exceeding 200mph have been observed. Jupiter is also undergoing massive atmospheric changes causing her plasma to appear much more highly charged and her rings to become brighter and brighter.

## MARS

Mars climate change is happening so fast that the planet may soon lose her southern ice cap completely. Although Mars changes are similar to those being experienced hereon Earth she is experiencing a much greater increase in storm activity and very strong winds ~ not yet seen on our planet.

# The Goddess VENUS

Venus is believed to have developed a "runaway greenhouse effect." Quite recently, she has also developed a new Red Spot and a red shift...both of which are completely mysterious to the scientific community.

# MERCURY

Mysterious Mercury is experiencing changes of such dynamics that she has been nicknamed **The Blooming Cosmic Rose**. The mysterious planet has also recently developed a "tail" ~ yes, a TAIL! Her new tail streams away from her surface in the opposite direction from the Sun, and can actually be seen with the naked eye on a clear night. She has also been showing an increase in both the frequency and intensity of magnetic tornadoes, causing scientists to conclude that Mercury is also in the midst of HYPER-DIMENSIONAL CHANGE.  It appears that she may currently be in a much higher, energetically-charged part of the galaxy than the rest of our planets.

## And ~ What about the Sun?

As stated previously, our Sun has been acting very "out of sorts" over the past couple of years.  In fact, she stayed at solar minimum so long that scientists were beginning to wonder if she'd ever wake up!  Then, quite mysteriously, the Sun went from having the lowest activity in 100 years to displaying a large number of increasingly strong flares  in early 2010 ~ **& ending with two M class flares!**

# ...All in All...

Our entire Solar System is within the midst of EXTREME CLIMATE and ATMOSPHERIC CHANGE ~ and the change is so dynamic that scientists refer to it as HYPER-DIMENSIONAL in nature!

**Is this what Dr. Agnew was referring to when he concluded:**

**"All the signs are already here!"**

# Artistic Rendering of the Surfaces of other Planets

# XIII

# Magical Mystery Tour

# MAGICAL MYSTERY TOUR

## Solar System in Hyper-Dimensional Change

O ver the past few years scientists and
astrophysicists have added oddly metaphysical
touches to their scientific explanations.  Believe it or
not, our ability to understand the magnificent universe
in which we live has been greatly enhanced through
the study of quantum physics and subatomic
particles!

## 2012~The Galactic Center

In other words, through the study and experimentation of the TINIEST of TINY subatomic particles ~ and their electrons, neutrons and protons ~ scientists have, quite recently, been able to add some amazing new knowledge to their understanding of the "mystery of life" within our universe! Through the manipulation of protons in lab experiments physicists have found that protons are not only be affected by ~ but also react. (in intelligent ways) to ~ the addition of different factors and stimulus!

A Magnificent View of the

## Galactic Center

or

## DARK  RIFT

As Seen from

Northern Arizona


Even more amazing is the fact that these
subatomic particles react to unseen factors such as
the INTENTION of the person performing the
experiment.  It has been PROVEN that the intention
of the of the individual running the experiment creates

247

an immediate affect upon the action of the particle...giving us an insight into such mysteries as psychic ability, clairvoyance and even the art of magic!

Perhaps even more importantly, sub-atomic and quantum physicists have recently proven that every atom ~ not only taken from a human body, but from any living thing ~ breaks down into smaller and smaller versions of itself until it finally reaches it's perfect state...a particle of PURE LIGHT! This may not sound extraordinary at first, but what it means is that YOU are truly are a "Magnificent Being of Light." YOU are much, much MORE than just the exterior shell of your human body. YOU ~ as stated so long ago by Einstein ~ are made of pure ENERGY of LIGHT that can neither "be created, nor destroyed!" The implications of this one fact alone are so extensive as to deserves a complete writing unto itself.

With this added knowledge, today's scientific community has almost unanimously accepted that fact that everything within our galaxy and universe is connected via a great electrical grid or giant MATRIX. Every living thing, be it plant, animal, human ~ or even rock and crystal ~ is constantly emitting and receiving electrically charged particles. And, this is a natural part of life that all living things have been totally unaware of! Yet, it is these positive and negative emissions that connect all of us and everything around us to the GREAT MATRIX. And...what is most amazing is that all of us ~ unknowingly ~ have always had the ability to create cause and affect results within this matrix.

With that in mind ~ and not taking into account any information with respect to 2012 or our current galactic position ~ it is important to note that astrophysicists currently believe that we are entering into a much more highly-charged part of our galaxy. In other words the electrically-charged, galactic energy all around us is increasing to such an extent as to be considered **hyper-dimensional in nature...and all of the planets within our solar system are presently experiencing changes that are part of this galactic Hyper-Dimensional shift.** Our planet Earth is not alone.~ every planet within our solar system, including the Sun itself ~ is currently experiencing enormous atmospheric and environmental changes... including global warming.

As we approach the Galactic Center of our Milky Way Galaxy it stands to reason that our planet would not be the only one affected by the mysterious forces we are approaching. In fact the furthest planets from the Sun would be the first to encounter change as they each cross the Galactic Center in succession. Making our way from the furthest planets from the SUN ~ Pluto and Neptune ~ down to planet Earth, Venus, Mercury, and finally... the Sun. In order to evaluate the changes being experienced within our solar system as each planet approaches the Galactic Center region, we will begin our tour at the farthest planet from the Sun – Pluto.  If our theory is correct we should see the affects of the Galactic Center and it's dynamic energy forces being felt the strongest at the outer edges of our solar system.

# Planet Pluto

Pluto, the furthest planet from the Sun, is also experiencing global warming...a fact that is not understood by the scientific community.  Recently-taken Hubble images reveal Pluto to be a complex-looking world variegated with white, dark-orange and charcoal-black terrain.  We have also learned that Pluto is not simply a ball of ice and rock, but a dynamic world that undergoes dramatic atmospheric changes.  In fact, ground-based observations taken in 1988 and 2002 show that the mass of Pluto's atmosphere has doubled over this time period.  Recently, scientists have noted that Pluto has experienced an alarming 300% increase in it's atmospheric pressure.  Pluto has also shown a dramatic change in color within two years (2000-2002) ~ during which time the planet has become significantly redder, while it's illuminated northern hemispheres is getting significantly brighter.

# Planet Neptune

What would it be like to awaken to a typical morning on Neptune?  Well, for one thing,  you can guarantee you will encounter a very windy and stormy day.

Neptune is known for her weird and violent weather, with massive storm systems and ferocious winds.  Be prepared for a weirdly stormy and quite windy day.  Windy to say the least!  Winds on Neptune average 900mph or 2,000km.

## The Rare Beauty of

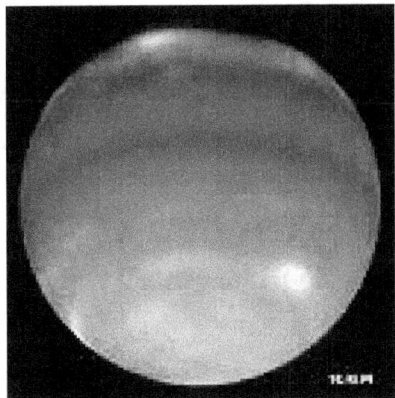

## Planet Neptune

The day will always be quite cool...and, oh yeah, don't bother with the sunscreen!  The sun, as seen from Neptune, is 900 times dimmer than it is as seen from planet Earth.  Seasons do change slightly, but each season lasts about 40 years because it takes 168 Earth years for Neptune to complete one Neptune year - or one complete revolution around the Sun.  But, if you happen to be a moon person, you'll most certainly be in heaven!  Neptune currently has 13 moons in revolution around itself.

And...most certainly your favorite feature of life on Neptune would be a tiny cloud by the name of Scooter. Scooter is a very small, irregular white cloud that literally "zzzzzzzzziiippppppsss" around Neptune every 16 or so hours.  Scientists have absolutely NO CLUE on this one...Scooter and her true nature remain an unsolved mystery!

Neptune's name came from ancient Roman mythology where she has always reigned as "Goddess of the Sea." And, as her name reveals, she is known for her very rich, vibrant and vividly blue color.  She is the 4th largest planet (in diameter) within our solar system, and is comprised of various ices and rocks.  Neptune has also been known for her Great Dark Spot, once her most prominent feature, which was about the size of ½ the diameter of Earth. The spot was located in the southern hemisphere and characteristically traveled westward across the planet, as it was pulled along by 700mph winds.  The Great

Dark Spot, however, has very recently disappeared! Neptune now has a new, very small dark spot in her northern hemisphere.

Neptune is also known for her many rings. These are seen as faint arcs but are actually complete rings with areas of bright clumps that can be seen as visible arc formations. Neptune currently has four rings: Adams, Galatea, Leverna and Galle. Adams consists of three separate arcs named Liberty, Equality and Fraternity; and Leverna is composed of two arcs named Lassell and Aargo.

Observations of Neptune made during a six-year period with NASA's Hubble Space telescope ~ by scientists from the University of Wisconsin-Madison and NASA's Jet Propulsion Laboratory~ show that the planet is exhibiting a significant and progressive increase in brightness. The observed changes show a distinctive increase in cloudiness, with a growing band of clouds in the southern hemisphere. The cloud formations are not only increasing in amount but also in brightness, and result in an increasing brightening of the banded cloud features that are distinctive of the planet.

"Neptune's cloud bands have been getting wider and brighter," according to Lawrence A. Sromovsky, a senior scientist as UW-Madison's Space Science and Engineering Center, and also a leading authority on Neptune's atmosphere. Neptune has been getting gradually brighter since 1980, but 2002 images from Hubble show that Neptune is much brighter than it was in 1996 and 1998. The increase in cloud activity first began in 1998 and increased dramatically in 2002.

# Planet Uranus

The rings of both Jupiter and Saturn seem to be changing...but Uranus is the leader! It's rings look completely different than they did 21 years ago when they were mapped by Voyager 2. The innermost ring is getting much closer to the planet and the brightness of various other rings is changing abruptly.

Uranus is the seventh planet from the Sun and the third largest with a diameter of about 51,000 km. Uranus is also known to be the most featureless planet to have been closely observed. So far, only a few icy clouds of methane have been observed. Uranus is unique amongst the planets in that her axis of rotation lies close to her orbital plane. As a result of her strongly tilted rotational axis, Uranus rolls "on her side" along her orbital path around the Sun, whereas other planets spin more or less upright. In other words, Uranus literally rotates sideways or tipped over onto her side.

Uranus is encircled by 11 moons that consist of rocks interspersed with dust lanes. The rings contain some of the darkest matter in the solar system and are extremely narrow, making them difficult to detect. Nine of them are less than 10 km wide, whereas most of Saturn's ring are thousands of km in width. Uranus has 15 known moons, all of which are icy and most of which are further out than the rings. The 10 inner moons are small and dark with diameters less than 150 km, and the 5 outer moons are larger with diameters between 470 km and 1500 km. The outer moons also have a wide variety of surface features. Her moon named Miranda has the most varied surface, with cratered areas broken up by huge ridges and cliffs up to 20 km high.

The atmosphere of Uranus is composed primarily of hydrogen and helium. It's inner regions are believed to consist of an "ocean" composed of ammonia, water and methane. The atmosphere then makes a gradual transition ~ without a clear boundary ~ into a gaseous atmosphere dominated by hydrogen and helium. Due to these differences, many astronomers group Uranus and Neptune into their own separate category ~ called the "Ice Giants." Uranus is very similar to Neptune in color, but instead of being a brilliant sapphire blue she has a more calming aquamarine appearance.

## Uranus ~ Recent CLIMATE CHANGE

Two billion miles away in the atmosphere of Uranus, a dark vortex large enough to engulf two-thirds of the United States has recently appeared. Measuring

1,100 by 1,900 miles in size, the Hubble telescope captured images of the phenomenon. Astronomers believe it to be a huge storm. Previous Hubble images of Uranus taken over the last decade have shown no dark spot, leading astronomers to believe that the disturbance has formed only very recently.

While rare on Uranus, dark spots have been frequently observed on Neptune. It was believed that while Uranus is similar in size and atmospheric composition to Neptune, it did not have such an active atmosphere. Lawrence Sromovsky, who led the team that took the pictures said that recently, however, "Uranus's atmosphere has shown an unusual increase in activity."

"People think that Uranus is relatively inactive, but these new images show that Uranus is definitely changing, and perhaps dramatically," according to Imke de Pater, Professor of Astronomy at the University of California, Berkeley. Professor Imke goes on to say, "What is causing it no one knows for sure...only time will tell."

# Planet Saturn

Perhaps the most dramatic evidence of some sort of climactic change on Saturn is the development of an enormous new ring!  Never before seen...the ring is way too large to miss...as it is now the largest of all Saturn's rings and extends downwards 170 MILLION MILES!

The ring was discovered recently by NASA's Spitzer Space Telescope and is equivalent to 20 times the diameter of the planet.  It would take about one billion Earths stacked together to fill the ring, or 300 Saturns lined up side-by-side.  One of Saturn's 49 moons, and one of it's farthest moons – Phoebe – circles within the newly found ring and may be the source of it's material.

# Planet Jupiter

Recently images of Jupiter show that the planet is clearly undergoing dramatic atmospheric changes that have never been seen before with the keen "eye" of NASA's Hubble Space Telescope.  This supports the idea that the planet is in the midst of **VIOLENT global climate change**!  Violent masses of storms with winds up to 200 mph have been observed.  And...even more startling...the turbulence and storms first seen on Jupiter over two years ago are still raging!

Both the Hubble and the Keck Observatories also reveal images showing that the bland, quiet band surrounding Jupiter's Great Red Spot - as observed only one year ago – has changed to an area of incredible turbulence on both sides of the spot. In addition, Jupiter appears to be experiencing plasma changes, with it's plasma becoming much more highly charged and it's rings getting brighter.

Jupiter has only 3 confirmed moons, and 55 other moons considered to be "irregular satellites" - plus one recently discovered irregular satellite that has not been named yet, and another whose orbit has not yet been established.

One of Jupiter's moons is known to be "one of the most exotic places in our solar system." Her moon named "Io", is the most volcanic body known to exist within our solar system...with lava flows, lava lakes, and giant calderas covering it's sulfurous landscape. It also shows billowing volcanic geysers spewing plumes up to over 500km high, and mountains much taller than those on Earth – some reaching the heights of 52,000 feet or 16 kilometers.

# Planet Mars

   "Mars is being hit by rapid climate change that is happening so fast the "Red Planet" could very soon lose it's southern ice cap," according to writer Jonathan Leakey. Scientists from NASA say that Mars has warmed by about 0.5C since the 1970s, and is experiencing similar changes to those seen on Earth over approximately the same time period. The mechanisms at work on Mars, however, appear to be a bit different than those on Earth. One researcher, Lori Fenton, believes that variations in radiation and temperature across the surface of Mars are generating extremely strong winds, not yet seen on our planet.

One thing is for certain...Mars is not a dead planet! According to James Head of Brown University, "Mars is now undergoing climate changes that are even more pronounced than those on Earth." But, like the Earth, Mars is also experiencing an increase in storm activity, and her ice caps are also melting.

# Planet VENUS

## Cool Green Goddess

Venus is the second-closest planet to the Sun and is named after the Roman "Goddess of Love and Beauty." After our moon, she is the brightest natural object in the night sky...reaching a magnitude of 4.6...and making her bright enough to cast shadows. Venus reaches her maximum brightness shortly before sunrise and after sunset, and is often called the "Morning Star" and the "Evening Star."

Although Venus is classified as a terrestrial planet, and sometimes called Earth's "Sister Planet" because of her similarity in size, gravity and bulk, scientists have no idea as to what the surface of the planet actually looks like. This is due to the fact that Venus is always covered by an opaque layer of highly reflective clouds which prevent her surface from being seen from space within visible light. In addition, Venus has no magnetic field, and the atmospheric pressure on her surface is enormous ~ 92 times that of the Earth. Venus is also unusual because she has no moons or other orbiting bodies. The only bodies orbiting Venus are the ones we have placed there such as NASA's Magellan Radar Satellite.

The Cool Green Goddess is believed by physicists to be the victim of a "runaway greenhouse effect." This occurs when there is positive feedback to the evaporation of all greenhouse gasses causing them to dissipate into the atmosphere. Quite recently, Venus has also developed a new "red shift" and small red spot...both of which are puzzling mysteries to scientists.

# Metallic Mysterious
# MERCURY

NASA's space probe recently revealed that the mysterious planet is experiencing some highly dynamic changes...resulting in scientists nicknaming her our "Blooming Cosmic Rose."  Mercury is the innermost planet to the Sun, and one of only three planets in our solar system that is a perfect sphere ~ the others being Venus and Pluto.  Mercury has a

## MYSTERIOUS MERCURY

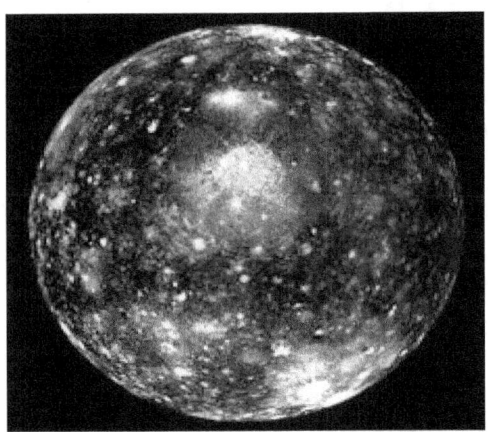

very thin atmosphere...in fact, it's atmosphere is nearly a vacuum.  Her surface is filled with moon-like craters and large, flat plains.  The planet has no weather to speak of, but it does have wild fluctuations in temperatures, and is known for having the largest

263

temperature spread of any planet in our solar system.

The sunny side of the planet may reach temperatures of 750 to 800 degrees F, while night-time temperatures plummet to nearly 330 degrees F. Since Mercury has no atmosphere with which to scatter the sunlight, the sky always appears to be black...very strange when taken into account that the Sun ~ as seen from Mercury ~ would be over twice as large as observed from planet Earth.

Photos taken from both the Mariner 10 and recent photos from the Mercury Messenger reveal Mercury to have a cratered surface marked with cliffs and ridges very much like our Moon. Lava flows have also been discovered, and complex radar obser-vations have found evidence of icy waters on her shady side.

But what types of changes are presently taking place on our neighboring, innermost planet? Observations from NASA's Space Probe reveal that Mercury has quite recently developed a "tail!" Yes...a tail...that streams away from the planet in the opposite direction from the Sun. Mercury has also been noted to have a recent increase in both the frequency and intensity of what scientists call "magnetic tornadoes"...leaving physicists to determine that "Mysterious Mercury" is also in the midst of hyper dimensional changes. According to physicists, the planet appears to be in a much higher, energetically-charged part of our galaxy...and that is why she is experiencing such extreme  hyper-dimensional change.

Mercury's magnetic field is strong enough to deflect the solar wind around the planet, creating a magnetosphere.  The planet's magnetosphere, though small enough to fit within the Earth, is strong enough to trap solar wind plasma.  This contributes to the space weathering of the planet's surface. Observations taken by the Mariner 10 Spacecraft detected this low energy plasma in the magneto-sphere of the planet's night side.  Bursts of energetic particles were also detected in the planet's "magnetotail" ~ which indicates an even more dynamic quality to the her magnetosphere.

# Sunrise on Mysterious Mercury

During its second flyby of the planet on October 6, 2008, MESSENGER discovered that Mercury's magnetic field can be extremely "leaky." The spacecraft encountered "magnetic tornadoes" ~ twisted bundles of magnetic fields connecting the planetary magnetic field to interplanetary space. These "tornadoes" were up to 800 km wide or a third of the radius of the planet. According to scientists they form when magnetic fields carried by the solar wind connect with Mercury's magnetic field. As the solar wind blows past Mercury, these joined magnetic fields are carried with it and twist up into vortex-like structures. These twisted magnetic "flux tubes" form open windows in the planet's magnetic shield through which the solar wind may enter and directly impact Mercury's surface.

The process of linking planetary and interplanetary magnetic fields is common throughout the cosmos; however, Mercury's reconnection rate is 10 times greater than that of all other planets in our solar system.

The MYSTERIOUS part of all this is that the Sun and it's proximity to Mercury only accounts for about one-third of Mercury's dynamic electromagnetic field...leaving the scientific community puzzled about where the remaining two-thirds of this phenomenon are coming from?

# XIV

## Perpetuum

## Vernnum

Is Eternal Spring

Once Again

Upon our Horizon

## January 2011
# Earth's Axis Shifts by 3-4%

According to astrophysicists this is a HUGE change since a Pole Shift would probably occur at a 10% shift. Beginning around 12/21/12 scientists began seeing very strange changes to various aspects of our magnetic field, along with other anomalies. They concluded that our planet had shifted on it's axis by around 4%. REASON: Unknown.

**12/21/10:** Earth's Magnetic Field went "off" for a brief period causing major disruption to air travel, etc. Then, a few days later, it suddenly switched back "on" with all sensors simultaneously measuring 100%.

**01/04/11:** All compass readings are off by 4 degrees North.

**01/06/11:** Astronomers, astrophysicists and astrologers all note that our Zodiac is ahead by 30 days ~ an unprecedented event meaning that we are accelerating in our Precessional Cycle.

**01/14/11:** All compasses measuring due North correctly again.

**01/11/11:** In Greenland the Midnight Sun rose 2 days early for the first time in recorded history. Mysteriously, it rose at 11:11 am.

# Coming Polar Shift

**Date: 01.10.09**

**Host: George Knapp – Coast-to-Coast AM**

**Guest:  Brent Miller -  The Horizon Project**

**The following is a transcript of this radio interview.**

**George Knapp** – Guest Host for Coast-to-Coast AM with George Noory - welcomed researcher **Brent Miller,** of The Horizon Project, who warned of mounting evidence that Earth is due for a polar shift which will end civilization as we know it. "It could happen tomorrow, it could happen 50 years from now, but all the evidence that we've been able to collect indicates that the next one is about to happen," Miller cautioned of the global catastrophe that he sees as imminent.

Describing the cataclysmic changes which would befall the planet during such an event, Miller said that as the poles move, the equator will be upended as well.

According to him, this will result in widespread geographical changes due to wild gravitational shifts. Noting the preponderance of water on the planet, he said that "during this sloshing about, as Earth wobbles and reorients itself, what happens is that the waters literally decimate all the coastal areas around the planet." Painting a grim picture of a post-shift world, Miller estimated that the event would "wipe out four billion people within six months and then a "Mad Max" scenario would ensue upon our planet."

Miller explained that the Horizon Project based their forecast for a polar shift on a growing number of anomalies around the world. He cited weather oddities, such as snowfall in Baghdad for the first time in over 100 years and the increase in earthquakes in unlikely locations. According to him, off-planet clues, like the increase in solar activity, also lend credence to a polar shift scenario occurring in the future.

While he suspected that the worlds' governments know of this impending disaster, he was doubtful of any official announcement, saying "I think we are given all the warning we're going to be given. We are currently seeing all of the signs."

**Website:** TheHorizonProject.com

**Video:**    The Horizon Project - Bracing for Tomorrow

# XV

## Just
## Plain
## Weird

# UPDATE: January 4th, 2011

Just 4 days into 2011…masses of blackbirds began falling from the sky in central Arkansas. Arkansas awoke Saturday to find thousands of dead birds across a 1.5-square-mile area. The estimate was raised from a thousand dead red-winged blackbirds to about 5,000 by Monday. The birds were the second mass wildlife death in Arkansas in recent days. Last week, about 100,000 dead and dying drum fish washed up along a 20-mile stretch of the Arkansas River. Mass wildlife deaths have also been recently reported in the states of Illinois, Indiana, Florida, Kentucky, Louisiana, Maryland, Tennessee, Texas and Tucson, Arizona.

And, unfortunately, such wildlife deaths have recently been reported to be happening worldwide The Star reported on January 5th that more than 40,000 dead Velvet swimming crabs appeared on the Thanet shoreline in England. Dead starfish, lobsters, sponges and anemones were found. in Brazil after 100 tons of fish turned up dead off the coast of Paraná. And, The bodies of thousands of turtledoves rained down from the sky in Faenza, Italy.

Within seven days, by January 11th of 2011, mass deaths of birds and fish had been reported from around the world including reports from China, Haiti, Germany, Japan, Quebec, Sweden, the UK, Viet Nam, Whales and New Zealand.

In Haiti: Authorities probed dead fish in a Haitian lake.

In Australia: Dead fish clogged a lake at an airport.

In Italy: Dead fish, clams and crabs stretched along a 2-mile length of coastline.

In New Zealand: Hundreds of snapper were found dead along stretches of beaches.

**Biologist have not found a cause of death at this time.**

# STRANGE...or...
# Just Plain WEIRD

Over the past few years the scientific community has stumbled upon a number of anomalies that are just **STRANGE ENOUGH** to come from a science-fiction thriller or a Steven King novel! The WEIRDEST of WEIRD of these EARTH anomalies are briefly outlined for you below.

## 2008

– **EARTH's ICE CAPS become OCEANS:**
On November 10th of 2008 the North Arctic Polar Region became an island for the first time in 125,000 years ~ and created a Northwest Passage shipping route, a new global shortcut, within just a few days.

## 2009

– **NEW OCEAN Appears OVERNIGHT:**
Last year an enormous trench opened in Ethiopia within a matter of hours and literally before the eyes of a group of geologists who just happened to be on site at the time. The trench is continuing to enlarge and fill with sea water and it is believed it will soon become a passage between two oceans.

## –    NASA's VOYAGER shows Solar System at Edge of InterStellar Cloud:

On December 26th of 2009 NASA reported that their Voyager had discovered that our solar system is very close to an unexpected interstellar cloud...this cloud or FLUFF is held at bay just beyond the edge of our solar system by our sun's magnetic field or heliosphere – a magnetic bubble more than 10 billion kilometers wide that acts as a shield and protects the inner solar system from cosmic rays and interstellar clouds from outside our galaxy. Scientists are concerned that the very strong magnetic field of this FLUFF cloud could compress our heliosphere even more than it is compressed now...possibly affecting Earth's terrestrial climate. Could this enormous cloud be the outer edge of the Dark Rift, our Galactic Center?

## –    FEMA CAMPS Preparing Across The USA

–    Many states have recently set up enormous emergency FEMA camps as noticed and reported by many citizens across the country, but not made known to the general public for obvious reasons.

## –    SATURN Showing Dramatic CHANGES:

On November 12th of 2008 Saturn mysteriously developed a new northern aurora that was photographed by NASA's Cassini spacecraft and the University of Arizona.

## – **MAGNETIC PORTALS DISCOVERED CONNECTING SUN & EARTH:**

On October 30th of 2008 NASA discovered magnetic portals that connect the Sun and the Earth. According to NASA scientists during Flux Transfer Events (FTE) the Earth's magnetic field presses against the sun's magnetic field every eight minutes. The two fields then briefly merge forming a portal through which particles can flow. These magnetic cylindrical portals tend to from above the Earth at the equator and then roll over the planet to the opposite pole. In winter the FTEs roll over the north pole and in summer they roll over the south pole.

## – **COSMIC RAYS BOMBARD EARTH:**

In November of 2008 an international team of research scientists discovered a puzzling amount of high-energy electrons or cosmic rays bombarding the Earth from deep space. The source is mysteriously unknown but scientists believe it must be close to the solar system and could be made of dark matter. Is it perhaps coincidental to note that the galactic center is home to what scientists call a massive "black hole" which they believe is made of dark matter...and that we are approaching the galactic center with our alignment on the 21st of December in 2012?

## – TWO MORE HOT SPOTS BOMARD EARTH with COSMIC RAYS:

On December 6th of 2008 in international team of researchers at Los Alamos National Laboratory's Milagro Observatory discovered two distinct hot spots (red regions) that appear to be bombarding Earth with an excess of cosmic rays ...the source is near the Orion constellation.

## 2010

## – ROGUE WAVES are INCREASING in FREQUENCY throughout all of our oceans:

Rogue waves were first noticed by satellite imagery a couple of years ago. A rogue wave is one single and enormous wave – hundreds of feet high – that literally appears out of nowhere somewhere out in the depths of the oceans and races around the planet at unbelievably high speeds. Such waves have never impacted a shoreline, but they have been a concern to shipping vessels and cruise ships since they are tall enough to literally devour them upon impact.

## -COSMIC DUST is HERE!

As we make our approach into the Galactic Center the gravitational and magnetic forces at play from the center of the galaxy are so unbelievably powerful that they literally pull and suck everything not stable enough to maintain it's own orbit into the galactic center ~ much like an enormous vacuum. As a result this area is always filled with cosmic dust and other space debris such as space rocks, asteroids, meteors, etc...and also why the center area of our galaxy is always seen as dark and cloudy and therefore named "The Dark Rift."

Cosmic dust was first observed entering into our solar system in 1993 by two spacecraft observatories. However, the amount of cosmic dust seeping through has tripled. within recent years. It is seen as a misty pink or salmon colored haze. There was a recent event in New Mexico where areas where people awoke to a mysterious layer of dust – everywhere – and the dust was so thick that it literally covered everything, much like a cosmic snowstorm! People were forced to clean off their windshields before they were even able to drive their cars. Alarmingly, this "dust" was of the same composition as the dust observed within the ice core samples previously discussed.

## -CO2 is the HIGHEST it has been for the past 20 MILLION YEARS...and STILL INCREASING!

CO2 is currently at 370 ppm. Deep ice core drilling at the Antarctica and the Russian Vostok Base provide scientists with data for CO2 trapped within the ice going back 420,000 years...and providing evidence that this level of CO2 is at an "unprecedented" high never before recorded on this planet.

## –  COSMIC DEBRIS at RECORD HIGH:

The Earth is presently being visited by an enormously large, unprecedented number of comets, asteroids and meteors on a monthly basis...and has been lucky enough to avoid a number of "almost" impacts within the first few months of 2010 alone!

## –  COLLIDING COMETS:

On January 29th of 2010 NASA's Space Hubble captured a never before seen event --- a head-on collision between two comets 90 mm (150 km) from Earth!

## –  EARTHQUAKES are Increasing in Intensity:

On February 27th of 2010 an 8.8 magnitude earthquake hit Chile.  The quake was 500 times stronger than the Haiti quake of two years ago and tsunami warnings went out as far away as the Hawaiian Islands.

## –    EARTH's AXIS makes a  SLIGHT CHANGE:

The Chilean earthquake was so powerful as to have made a slight change in our planet's axis of rotation ~ which  was already tilted and "wobbling."

## –    EARTH'S MAGNETIC FIELD has recently become TWISTED and WEAK:

Scientists have noticed that our planet's electromagnetic field has been losing strength and becoming extremely weak and scattered within the past few years. Recently it has also taken on a never before seen "twisted" configuration....never before seen and remaining a mystery to the scientific community.

## –    Earth's North and South GEOGRAPHIC POLES Slowly Shifting:

Scientists have noted that both our North and South geographic poles have shown a recent, very slight, shift.

## – PHYSICISTS NOTE that TIME is ACTING STRANGELY:

Astrophysicists have recently noted that "time" appears to be "speeding up."

## – SUN Acting Out of Character:

Our Sun recently completed a solar cycle but remained very quiet for an unprecedented amount of time. It "awoke" in January of 2010 after which scientists noticed that there had been a change in our heliosphere for the first time in recorded history. The heliosphere is a protective layer created by the Sun's rays that acts as an enormous shield, protecting the Earth and other planets by literally repelling comets, asteroids, lethal rays and cosmic dust from entering into our solar system. Physicists believe that during it's long quiet spell our Sun failed to completely close off it's heliosphere – yet another unprecedented event never before seen by scientists and astrophysicists.

## – Scientists Declare that EARTH is now WELL INTO her 6TH Extinction!

The scientific community has declared that we are presently within the midst of our planet's 6th Extinction due to the increasingly high rate of extinction of both plant and animal species on the planet.

## –    TWO COMETS PLUNGE into SUN;

**January 2010:** Two comets plunged directly into the Sun! On January 4th the first comet disintegrated upon approach, and on January 22th the second comet vaporized upon approach. Once again, this was an unprecedented event...the Sun normally repels all comets, meteors and asteroids with it's heliosphere.

## –    EARTH has CLOSE CALL with Direct  ASTEROID IMPACT:

On January 13th of 2010 an asteroid (2010AL30) barely missed a direct impact with our planet ~ approaching very close at 0.3 Lunar Distance.

## –    The SUN AWAKENS with Five SOLAR FLARES & Two M-Class Flares:

**In January & February 2010** our Sun ended it's abnormally long quiet period with five solar flare bursts within two days from January 18th through January 20th of 2010 ~ and continued to crackle with numerous M Class flares on February 8th through February 10th of 2010.

## –   MYSTERIOUS DARK DISK Covers BRIGHT STAR:

On January 20th of 2010 astronomers found a mysterious dark disk moving in front of Epsilon Aurigae...an extremely bright star that had been losing it's brightness for the first time in 27 years. Epsilon Aurigae is a bright star that has puzzled astronomers for 175 years.

## –   The SUN Develops a One-MILLION-Kilometer Long FILAMENT

**In FEBRUARY 2010** the Sun continued to amaze NASA scientists on the 20th through the 22nd of February 2010 when one of their spacecraft discovered an extremely long filament projecting out of the Sun that stretched more than a million kilometers around the Sun's southeastern limb. Their concern was that similar filaments have been known to collapse, and when they hit the Sun's surface create a tremendous explosion call a "HyderFlare."

## –   MORE COMETS Plunge into Sun:

**On March 11th, 12th and 13th of 2010** four comets plunged DIRECTLY into the Sun.

## –　A Coronal Mass Ejection Heads for EARTH:

On March 16th of 2010 both a solar wind stream and a coronal mass ejection headed for Earth...resulting in a geomagnetic storm alert for March 17th and 18th.

## –　Astrophysicists Worry About Galactic SUPERWAVES Impacting Earth:

In early 2010 astrophysicists ponder the threat of Galactic Super Waves hitting Earth from the galactic center within the next few years.

## –　Scientists DISCOVER Solar System in Hyper-Dimensional Change:

All PLANETS within our Solar System are EXPERIENCING DYNAMIC CLIMATE and ATMOSPHERIC CHANGES, including Global Warming.

....**Things are definitely becoming very active  and strange**

**both on and around our planet...**

## .....STRANGE?........

Or

# JUST PLAIN WEIRD!

# XVI

## What will

## Happen

## ...or Not?

# The Petrified Forest in Northern Arizona

Northern Arizona is one of the few places on the planet that exhibits a large forest of Petrified Trees. Scientists estimate this forest was formed anywhere from 200 to 225 million years ago ~ which places the event around the timeline of the Great Permian Extinction, and around the time when Arizona was covered by a shallow ocean. The Permian Extinction also took at the time of our last passage through the Galactic Center. The genotype of these ancient trees became extinct at that time, but the once massive forest is estimated to have been filled with trees around 200 feet tall. Then, suddenly, the entire forest was literally flattened by an enormous wave that quickly washed over it.

Petrifaction can only occur when a living organism is covered very quickly by massive amounts of water, and then buried deeply underneath the weight of tons and tons of the water for millions of years. The wood then becomes crystallized ~ as can be seen in the above photo.  A massive tsunami?  Most likely caused by a geographical pole shift? Very possibly, and worthy of further investigation.

# What will Happen...

# or Not?

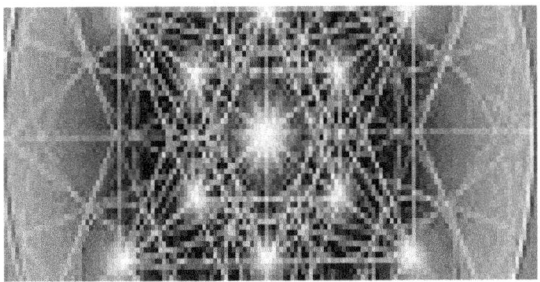

**The Great Matrix:**  Scientists have recently come up with a new theory that the entire universe is interconnected and held together by what can be visualized as an enormous, electrically-integrated grid or net-like structure ~ which they have named The Great Matrix.

# TIME WAVE ZERO

## 12:21:12:11:11:00:00

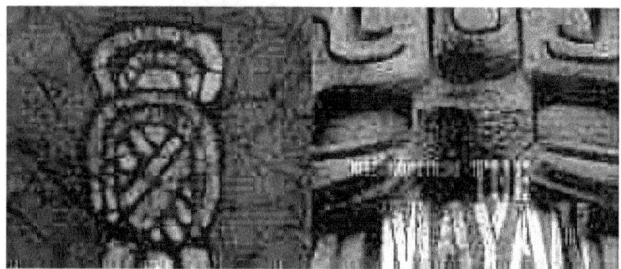

**The Mystery Left for Us by The Ancient Mayans**

It should be clear to everyone reading this book that we are in the midst of something very, very unique and unprecedented within the history of man on this planet.

What will happen on December 21st of 2012 at 11:11AM? Should we expect something of critical importance to take place even before that date? What do the most brilliant scientists theorize could or could not take place? What would it be like to experience a "day in the life of a complete polar flip?" Are there any areas of our planet that are considered to be "Safe Havens"? What parts of the country and the world are most at risk during this upcoming critical time?

# &

# ...Finally...

## How shall we PREPARE ourselves, our families, and our loved ones to meet with the 2012 alignment?

The answers to all of these questions will be addressed in the second & third books of **Trilogy 2012. The Trilogy** will also introduce readers with how to prepare for the spiritual and metaphysical aspects of the 2012 phenomenon ~ which is also critical in order to meet with and survive this **crossing of a Chasm of Millions of Years!**

# Book II: Trilogy 2012 includes:

How to  Prepare for 2012

A Survival Checklist

A Survival Guide to prepare for long-term
survival after the event

A listing of Safe Havens

&

An intriguing look at what it would be like to
live through A Day in the Life of a
Geographical Pole Shift.

**For updates on the release of Book
II & Book III:  please  e-mail:
caryn911@yahoo.com.**

# XVII

# The Beginning
## of

# the End?

## OR

# The Beginning of a

# New Beginning?

# SCIENCE

# NOT Superstition

# 12:21:12:11:11:00:00

## EARTH REACHES POINT ZERO

## &

## The "Finger of GOD"

As our Sun, Earth and the rest of our Solar System reach Point Zero at 11:11AM on the 21st day of December 2012 ~ scientists, physicists and astro- physicists all agree that we should be met with intense, centripetal and magnetic forces....which should continue during the hours it takes to pass through and across this plane of our Galactic Center.

This should become "The Question for the Ages" ~ because although the scientific community can theorize on what could happen...it is completely impossible to know for certain.  This is due to the fact that we are dealing with enormous forces coming from both within in Galactic Center and also from outside the galaxy itself.  And...as we have discussed...we know very little about the galactic center due to the fact that it is so dark and obscured at all times.  In the words of Dr. Brooks Agnew, "...we are not able to SEE the forces at play...but we can FEEL some of the forces as they pass through our atmosphere as different types of wave frequencies."

Our universe does not do anything by accident. There is a Grande Design to everything, everywhere. And, there is no such thing as coincidence within the "Designer's Plans."  So ~ it is imperative that we recognize and accept that it is most certainly our FATED DESTINY to BE HERE NOW! YOU were meant to BE HERE NOW...and like everything else, each one of our FATES is unique and personal and in the "Hands of God."

Back to the FACTS ~ there is some possibly BAD news, and there is some possibly GOOD news.  The BAD news is that the scientific community believes that the WORST thing that can happen during this critical time is that we would experience an immediate GEOGRAPHICAL POLE SHIFT ~ or possibly one or more POLE SHIFTS.  The Pole Shift is believed to take place "instantly" and will be over within a matter of minutes or hours. However...it can be devastating to the majority of the land, and life, on our planet.

The GOOD news is that during this critical alignment three of our planets ~ Pluto, Neptune and Saturn ~ come together to form what the ancient Mayans called "The Finger of God." Also meeting up with us at the Galactic Center is a large star called 4SGM and the Trifled M20 Nebulae.

## Other celestial anomalies taking place during the year 2012 are:

—      A RARE transit of Venus on June 5th and 6th and during which time Venus also conjuncts with the central star of the Pleiades Star System ~ a star named Alcyone.

—      There will be two MAJOR Solar Eclipses in the year 2012:

# RARE SOLAR ECLIPSES in 2012

**1)** May 20th -~ The Sun and Moon conjunct with the Pleiades Star System during their transits.

**2)** November 12th ~ The Sun and Moon conjunct with the Constellation Serpens during their transits. Serpens is known for representing Eternity.

# In Addition:

− Three of our Planets take position to form "The Finger of God."

− We come very close to Star GM.

− We come very close to the Trifled M20 Nebulae.

...It appears that **someone** has put a lot of time and effort into a GRANDE PRODUCTION to take place on the 21st of December 2012...one that holds many heavenly meanings for those whom seek...

# ~ FORTUNATELY~

This is one book that must

leave you to write  your own,

unique...and very personal

## ~CONCLUSION~

# Please look for Trilogy 2012 : Part II and III:

## Trilogy 2012 Book II

A Day in the Life of a Pole Shift

Safe Havens, How to Prepare & Survival Guide for 2012.

## Trilogy 2012 Book III

Metaphysical & Spiritual Preparedness

for Meeting 2012.

12:21:12:11:11:00:00                                          2012

"Look
   Look Up
    Look Around
           Look Behind
           Look Beyond
Then........Just Keep on Looking!

Look at the FACTS
   Look through Them
    Look Around Them
      Even Look through the Eyes of
      Someone Else

## BUT
   Whatever You Do
      NEVER Stop LOOKING!"

~~~

# AND

Remember Just ONE Thing

No Matter What You Find....
...KEEP on INVESTIGATING...

..        .NEVER STOP LOOKING...

INVESTIGATE  LIFE
&
...NEVER Forget...

# Knowledge
### is Your Only
# Power!

# REFERENCE INDEX

Universe – Ch 12

University of Arizona – Ch 11:10, 14

University of California, Berkeley – Ch 12

University of Chicago – Ch 11:11

University of New Hampshire – Ch 11:9

University of Wisconsin-Madison – Ch 12

Uranus – Ch 12

US Geological Survey – Ch 11:0

UW-Madison's Space Science & Engineering Center – Ch 12


Vatican – Ch 14

Vega – Ch 7

Vernum Pertetuum – Ch 4

Venus – Ch 12

Virgo Cluster – Ch 3, 8

VLA (Very Large Array Observatory) – Ch 11:11, 11:12

VLBA (Very Long Baseline Radio) – Ch 7

Volcanic Activity – Ch 11.6

Volcano Data Charts – Ch 11.7

Volcanoes – Ch 10, 11.7

Voyager Spacecraft – Ch 11.8, 14

# RARE CELESTIAL EVENTS
# in 2012

**January 3, 4** - Quadrantids Meteor Shower. The Quadrantids are an above average shower, with up to 40 meteors per hour at their peak. The shower approaches us from the constellation Bootes.

**March 20** - The Vernal Equinox IIII in the northern hemisphere at 05:14 UT. There will be equal amounts of day and night. This is also the first day of spring.

**April 15** - Saturn at Opposition. The ringed planet will be at its closest approach to Earth. Thin and its moons.

**April 21, 22** - Lyrids Meteor Shower. The Lyrids are an average shower, usually producing about 20 meteors per hour at their peak. These meteors can produce bright dust trails that last for several seconds. The shower usually peaks on April 21 & 22, although some meteors can be visible from April 16 - 25. Look for meteors radiating from the constellation of Lyra after midnight.

**May 5, 6** - Eta Aquarids Meteor Shower. The Eta Aquarids are light showers, usually producing about 10 meteors per hour.

**May 20** - Annular Solar Eclipse. The path of which will begin in southern China and move east through Japan, the northern Pacific Ocean, and into the western United States. A partial eclipse will be visible throughout parts of eastern Asia and most of North America. (NASA Map and Eclipse Information | NASA Eclipse Animation)

**June 6** - Transit of Venus Across the Sun. A partial transit can be seen in progress at sunset throughout most of North America, Central America, and western Europe. (NASA Transit Information | NASA Transit Map)

**June 20** - The Summer Solstice occurs in the northern hemisphere at 23:09 UT, when the Sun is at its highest point in the sky.

**July 28, 29** - Southern Delta Aquarids Meteor Shower. The Delta Aquarids can produce about 20 meteors per hour at their peak. The shower usually peaks on July 28 & 29, but some meteors can also be seen from July 18 - August 18. The radiant point for this shower will be in the constellation Aquarius. Best viewing is usually to the east after midnight.

**August 12, 13** - Perseids Meteor Shower. The Perseids is one of the best meteor showers to observe, producing up to 60 meteors per hour at their peak. The shower's peak usually occurs on August 13 & 14, but you may be able to see some meteors any time from July 23 - August 22. The radiant point for this shower will be in the constellation Perseus. Look to the northeast after midnight.

**August 24** - Neptune at Opposition.

**September 22** - The Autumnal Equinox will occur in the northern hemisphere at 14:49 UT. There will be equal amounts of day and night.

**October 21, 22** - Orionids Meteor Shower. The Orionids is an average shower producing about 20 meteors per hour at their peak. This shower usually peaks on the 21st, but it is highly irregular. A good show could be experienced on any morning from October 20 - 24, and some meteors may be seen any time from October 17 - 25. Best viewing will be to the east after midnight.

**November 13** - Total Solar Eclipse. The path of totality will only be visible in parts of extreme northern Australia and the southern Pacific Ocean. A partial eclipse will be visible in most parts of eastern Australia and New Zealand. (NASA Map and Eclipse Information | NASA Eclipse Animation.

**November 17, 18** - Leonids Meteor Shower. The Leonids is one of the better meteor showers to observe, producing an average of 40 meteors per hour at their peak. The shower itself has a cyclic peak year every 33 years where hundreds of meteors can be seen each hour. The last of these occurred in 2001. The shower usually peaks on November 17 & 18, but you may see some meteors from November  13 - 20. Look for the shower radiating from the constellation Leo after midnight.

**November 28** - Penumbral Lunar Eclipse. The eclipse will be visible throughout most of Europe, eastern Africa, Asia, Australia, the Pacific Ocean, and North America.

**December 3** - Jupiter at Opposition. The giant planet will be at its closest approach to Earth.

**December 13, 14** - Geminids Meteor Shower. Considered by many to be the best meteor shower in the heavens, the Geminids are known for producing up to 60 multicolored meteors per hour at their peak. The peak of the shower usually occurs around December 13 & 14, although some meteors should be visible from December 6 - 19. The radiant point for this shower will be in the constellation Gemini. Best viewing is usually to the east after midnight.

**December 21** - The Winter Solstice occurs in the northern hemisphere at 11:11 UT. The Sun is at its lowest point in the sky and it will be the shortest day of the year. This is also the first day of winter.

12:21:12:11:11:00:00                                    2012

12:21:12:11:11:00:00                              2012

326